CTBUH Awards
Best Tall Buildings

A Global Overview of 2013 Skyscrapers

Antony Wood, Steven Henry & Daniel Safarik

NEW YORK AND LONDON

NEW YORK AND LONDON

Bibliographic Reference:
Wood, A., Henry, S. & Safarik, D. (2014) *Best Tall Buildings: A Global Overview of 2013 Skyscrapers*. Council on Tall Buildings and Urban Habitat: Chicago.

Book Design & Layout: Marty Carver

First published 2014 by Routledge
2 Park Square, Milton Park, Abingdon, Oxon OX14 4RN

Simultaneously published in the USA and Canada by Routledge
711 Third Avenue, New York, NY 10017

Routledge is an imprint of the Taylor & Francis Group, an informa business

Published in conjunction with the Council on Tall Buildings and Urban Habitat (CTBUH) and the Illinois Institute of Technology

© 2014 Council on Tall Buildings and Urban Habitat

Printed and bound in the USA by Sheridan Books, Inc.

The right of The Council on Tall Buildings and Urban Habitat to be identified as author of this work has been asserted by it in accordance with sections 77 and 78 of the Copyright, Designs and Patents Act 1988.

All rights reserved. No part of this book may be reprinted or reproduced or utilized in any form or by any electronic, mechanical, or other means, now known or hereafter invented, including photocopying and recording, or in any information storage or retrieval system, without permission in writing from the publishers.

Trademark notice: Product or corporate names may be trademarks or registered trademarks, and are used only for identification and explanation without intent to infringe.

British Library Cataloguing in Publication Data
A catalogue record for this book is available from the British Library

Library of Congress Cataloging in Publication Data
A catalog record has been requested for this book

ISBN13 978-0-415-73717-3
ISSN 1948-1012

Council on Tall Buildings and Urban Habitat
S.R. Crown Hall
Illinois Institute of Technology
3360 South State Street
Chicago, IL 60616
Phone: +1 (312) 567-3487
Fax: +1 (312) 567-3820
Email: info@ctbuh.org
www.ctbuh.org

Printed and bound in the United States of America by Sheridan Books, Inc. (a Sheridan Group Company).

Acknowledgments

The CTBUH would like to thank all the organizations who submitted their projects for consideration in the 2013 awards program.

We would also like to thank our 2013 Awards Jury for volunteering their time and efforts in deliberating this year's winners.

About the CTBUH

The Council on Tall Buildings and Urban Habitat is the world's leading resource for professionals focused on the design, construction, and operation of tall buildings and future cities. A not-for-profit organization based at the Illinois Institute of Technology, Chicago, the group facilitates the exchange of the latest knowledge available on tall buildings around the world through events, publications, research, working groups, web resources, and its extensive network of international representatives. Its free database on tall buildings, The Skyscraper Center, is updated daily with detailed information, images, data, and news. The CTBUH also developed the international standards for measuring tall building height and is recognized as the arbiter for bestowing such designations as "The World's Tallest Building."

Contents

Foreword	6
Introduction	8
CTBUH Best Tall Building Awards Criteria	25

Best Tall Building Americas
Winner:
The Bow, *Calgary*	28

Finalists:
Devon Energy Center, *Oklahoma City*	34
Tree House Residence Hall, *Boston*	38

Nominees:
1214 Fifth Avenue, *New York*	42
Ann & Robert H. Lurie Children's Hospital, *Chicago*	44
LOVFT, *Santa Catarina*	46
Mercedes House, *New York*	48
Trump International Hotel & Tower, *Toronto*	50
Coast at Lakeshore East, *Chicago*	52
Helicon, *San Pedro Garza Garcia*	52
Pacifica Honolulu, *Honolulu*	53
Reforma 342, *Mexico City*	53
Rush University Medical Center Hospital, *Chicago*	54
Torre Begonias, *Lima*	54
Torre Paseo Colón 1, *San José*	55

Best Tall Building Asia & Australasia
Winner:
CCTV Headquarters, *Beijing*	58

Finalists:
C&D International Tower, *Xiamen*	64
PARKROYAL on Pickering, *Singapore*	68
Pearl River Tower, *Guangzhou*	72
Sliced Porosity Block, *Chengdu*	76

Nominees:
Brookfield Place, *Perth*	80
Hangzhou Civic Center, *Hangzhou*	82
Hysan Place, *Hong Kong*	84
International Finance Centre, *Seoul*	86
I Tower, *Incheon*	88
Japan Post Tower, *Tokyo*	90
NBF Osaki Building, *Tokyo*	92
Shenzhen Stock Exchange, *Shenzhen*	94
Shibuya Hikarie, *Tokyo*	96
Soul, *Gold Coast*	98
The Ellipse 360, *New Taipei City*	100
Zhengzhou Greenland Plaza, *Zhengzhou*	102
Alamanda Office Tower, *Jakarta*	104
ARK Hills Sengokuyama Mori Tower, *Tokyo*	104
City Tower Kobe Sannomiya, *Kobe*	105
Dolphin Plaza, *Hanoi*	105
Huarun Tower, *Chengdu*	106
Net Metropolis, *Manila*	106
Pyne, *Bangkok*	107
Reflection Jomtien Beach, *Pattaya*	107
Shenzhen Kerry Plaza Phase II, *Shenzhen*	108
Yixing Dongjiu, *Yixing*	108
Yokohama Mitsui Building, *Yokohama*	109

Best Tall Building Europe
Winner:
The Shard, *London*	112

Finalists:
ADAC Headquarters, *Munich*	118
New Babylon, *The Hague*	122
Tour Total, *Berlin*	126

Nominees:

Mercury City, *Moscow*	130
Unicredit Tower, *Milan*	132
No. 1 Great Marlborough Street, *Manchester*	134
Torre Unipol, *Bologna*	134
Varyap Meridian Block A, *Istanbul*	135

Best Tall Building Middle East & Africa

Winner:

Sowwah Square, *Abu Dhabi*	138

Finalists:

6 Remez Tower, *Tel Aviv*	144
Gate Towers, *Abu Dhabi*	148

Nominees:

JW Marriott Marquis, *Dubai*	152
Diplomat Commercial Office Tower, *Manama*	154
Faire Tower, *Ramat-Gan*	154
Frishman 46, *Tel Aviv*	155

10 Year & Innovation Awards

10 Year Award Winner:

30 St Mary Axe, *London*	158

Innovation Award Winners:

BSB Prefabricated Construction Method	164
KONE UltraRope	168

Innovation Award Finalists:

Megatruss Seismic Isolation Structure	172
Precast Concrete Façade	176
Rocker Façade Support System	180

Lifetime Achievement Awards

Lynn S. Beedle Award, *Henry N. Cobb*	186
Fazlur R. Khan Medal, *Clyde N. Baker, Jr.*	192
CTBUH 2013 Fellows	198

Awards & CTBUH Information

CTBUH 2013 Awards Jury	199
Review of Last Year's CTBUH 2012 Awards	200
Overview of All Past Winners	206
CTBUH Height Criteria	208
100 Tallest Buildings in the World	211

Index

Index of Buildings	216
Index of Companies	217
Image Credits	220
CTBUH Organizational Structure & Members	222

Foreword

Jeanne Gang, *2013 Awards Jury Chair*

It has been an honor to serve as Chair of the 2013 CTBUH Tall Building Awards. Taking time to review and compare built work from around the world offers deep insight into the state of our collective practices, priorities, and societies. As expected of an international jury composed of architects, engineers, and technology and sustainability experts, our discussions were wide-ranging, touching on the many facets of tall buildings and their development around the globe.

Yet what resonated most for me was that, for many cities, tall buildings have become the "new normal." Thousands of high-rise buildings already exist and form the urban context in hundreds of cities worldwide. There are 490 new buildings scaling 200 meters or higher currently under construction around the world, with 11 over 500 meters in height. Tall buildings are no longer rare exceptions to the rule; they are becoming the preferred modality of the city.

While building tall retains unique technical challenges once reserved for iconic corporate headquarters, skyscrapers are now being enlisted to provide everyday uses such as workspace, accommodation, commerce, healthcare, and even education. These programs are finding new ways to navigate the vertical dimension. And although building tall is not the only solution to society's growth issues, it is a necessary part of sensitive land use and urban design in the twenty-first century. Tall buildings are providing smart solutions for expanding populations in new and existing – and even historic – cities.

The proliferation of these tall buildings means that it is more urgent than ever before to strive toward better designed, better performing, and less polluting strategies. This year's entries demonstrate progress along these fronts, with vertical structures continuing to employ smarter, more sustainable solutions with more ambitious structures and forms. And while some of the buildings in this year's competition have been on our collective radar for a while, this year marked their formal completion and a great moment to recognize their contributions.

In Asia, the completion of the Pearl River Tower stresses the relevance of increased energy production and performance, while the Sliced Porosity Block demonstrates the importance of better urbanism, evidenced in its creation of public space where the towers meet the ground. Recently completed buildings in Europe, including the Tour Total and ADAC headquarters, signal a return to materiality, adding liveliness and color to the streetscapes and challenging the preponderance of purely glass façades that have long held sway. And in buildings such as The Bow in Calgary and the CCTV tower in Beijing, structure is again celebrated, and inventive conceptual ideas are articulated through form and program.

The great variety of technical and programmatic solutions presented in this year's entries indicates that maturation in tall building design is coinciding with exciting experimentation. The sheer quantity and breadth of these projects translated into stronger judging criteria and the need for winning entries to work on multiple levels. The projects recognized in this publication have all, in some way, contributed to the progressive evolution of tall buildings.

This evolution is perhaps most obvious in the entries we reviewed in this year's Innovation category, with several outstanding innovations hinting at the future of tall building construction. The Broad Sustainable Building (BSB) Prefabricated Construction Method, for example, captured the industry's attention in 2012 with the construction of a 30-story hotel building in just 15 days in Changsha, China – an innovation that has significant implications for future building. Similarly, the KONE UltraRope, a carbon-fiber hoisting technology whose weight and bending advantages effectively double the distance an elevator can travel in a single shaft, has the potential to significantly alter current barriers to height.

These and other research projects are vital to the continued progress of the tall building with regard to its sustainability and efficiency as well as its construction and operation. They epitomize the way innovative technologies work together to change our perception of what is possible within this typology and to inspire future invention.

It is with the utmost respect and admiration that I thank my fellow jurors Richard Cook, Nenjun Luo, Robert Okpala, David Scott, Karen Weigert, and Antony Wood as well as organizer Steven Henry at the CTBUH, for making the judging process so very enriching. Their contributions to our discussions on each of the many interesting projects demonstrate a deeply rooted commitment to moving architecture forward, and I am truly privileged to have stewarded this journey.

Introduction

Antony Wood, *CTBUH Executive Director*

Abu Dhabi, Beijing, Calgary, London. I wonder if, even ten years ago, we'd have predicted that the Best Tall Buildings in the World completed in 2013 would be based in those four cities, as opposed to the New Yorks, Dubais, or Shanghais of the world, where the center of gravity of tall buildings – certainly in terms of numbers – seems to have been focused.

This is, of course, further evidence of the biggest trend that has been happening in tall buildings over the past couple of decades; the simultaneous rise of skyscraper cities the world over. No longer concentrated in the US, or even Asia, we are seeing the vertical rise of cities on every continent and in virtually every region globally – increasingly so now, as the world starts to rebound from the financial crisis of five years ago.

Less apparent, but no less interesting, is the fact that, if we seek quality or innovation in tall building design and engineering – as is obviously the case with this book and the awards program that supports it – then it is often these "second-tier" vertical cities that hold the best examples, possibly enabled by their relative distance from the heated pace of construction in the megacities.

We believe that the four 2013 winning regional projects as profiled in these pages certainly do exhibit high levels of quality and innovation, but in different ways. The Bow in Calgary not only applies a steel diagrid to a curved skyscraper for the first time in North America; the building plan curves concavely into the predominant sun direction, symbolically and physically embracing the sun into the building, when the majority of buildings strive to block it out. Once inside, the building puts this solar gain to good use, especially during the cold northern winters, through a series of façade atria that act simultaneously as buffer zones and depositories for the excess heat built up in the office spaces beyond. The inclusion of three sky gardens within this façade atrium zone, and the embrace of a large public plaza at the ground plane created by the curve of the building, also assists with the social sustainability of the project.

Opposite: Americas winner The Bow, Calgary. The concave form is supported by a diagrid system and encloses façade atria and sky gardens.

Left: Asia & Australasia winner CCTV Headquarters, Beijing. A technically daring form creates a dramatic cantilever, helping make the CCTV Building an icon years before it actually completed construction.

By providing shared, semi-recreational spaces at height, where a sense of building community can develop, as well as a strong interface at the ground floor, where the public and building community meet, The Bow takes a bold step in a direction most tall buildings fail to address.

Some readers may be surprised to see the CCTV Building in Beijing winning the Asia & Australasia award this year. The building – already topped out and clad in 2008 – had such an iconic presence during the Beijing Olympics that many thought it complete several years ago. The reality, however, is that the building only became occupied and operational – an important part of the criteria for "completion" as defined by the CTBUH (see page 210) – in 2012, hence its inclusion in the awards this year. Unlike The Bow, where the innovations are primarily in environmental approach, the main innovations of CCTV are in form and internal planning. Though the building image has already been consumed millions of times and replicated on the pages of books, journals, and in city branding since it was first proposed almost a decade ago, I ask you to cast your mind back to the first time you saw the proposition of this looping, cantilevering "anti-skyscraper" (as the architect intended it initially). Has a form as stylistically simple, and yet technically daring as this ever been attempted before? Though we have, by now, grown somewhat desensitized to that form due to its extensive media coverage, the jury felt it important to recognize the achievements of the building, and the positive impact it has already had on Beijing. Beyond form, the building program is amazingly complex, yet skillfully contained, with a "public loop" route through the building that takes visitors direct from the subway, up past television studios and associated facilities, to the spectacular observation deck with views out over the city, and down through glass apertures in the floor to the public plaza below. Speaking personally, I can say that visiting the CCTV Building is an amazing experience.

The Shard in London is, in many ways, no less remarkable than CCTV – not least in the negotiation of the considerable planning hurdles its creators overcame to make a supertall (300 meter+) building possible in such a historic city as London. The innovations in this beautiful column of glass include the clever accommodation of the differing physical requirements of its program within the tapering form: the smaller floor plates support residential and hotel programing in the top half, while larger floor plates support offices in the lower half. The project is also remarkable for its integration with, and rejuvenation of, the major transport

hub that is London Bridge Railway and Underground station. Certainly from a US perspective, it is incredible to think that a building of this size could be realized while providing only 48 direct car parking spaces. This points clearly to a future in which tall buildings – in order to make sense of their sustainable integration in the city – need to be linked directly to major transport infrastructure. Ultimately, in considering The Shard, the mere fact that the Tallest Building in Europe now resides in London (at least until other planned projects in Moscow are completed), is incredible enough in itself. There are very few who would have thought that possible a couple of decades ago.

In the Middle East & Africa category, we see the winner, Sowwah Square, amidst the deserts of Abu Dhabi. Although likely the least well-known of the four regional winning projects, this project impressed the jury massively with its environmental approaches, in a region that has seen few examples of environmental consideration in high-rises. One can only imagine the discussions that were necessary to convince a client and the authorities of the benefits of an active, ventilated double-skin façade; despite the obvious insanity of building single-layer all-glass skins in an intense desert-solar environment. Sowwah Square has achieved innovation environmentally in a way that most projects throughout the world – and certainly in the Middle East – have not. And, in the process, quite a beautiful composition has been created, one that seems rooted to its context culturally as well as environmentally.

Thus, though there are certainly issues and criticisms that could be leveled against all of the winning projects, against the background of many relatively uninspiring, un-innovative tall buildings being built the world over, these four projects do strive to do something different, and are at least partially successful in that quest.

This book is increasingly becoming a global snapshot of tall building completion in any given year, beyond just awards profiling. As such, most of the projects included

Opposite Left: Europe winner The Shard, London. The building integrates into a major transportation hub, highlighting the importance of the need for connecting tall buildings to major transit infrastructure.

Opposite Right: Middle East & Africa winner Sowwah Square, Abu Dhabi. The ventilated double-skin façade is an environmental response not often seen (though desperately needed) in this intense desert environment.

Top: Americas Finalist Tree House Residence Hall, Boston. The unique façade is expressive of the program within, student housing.

Bottom: Asia & Australasia Finalist PARKROYAL on Pickering, Singapore. Large terraces bring dense vegetation up the height of the building.

in this publication have something special about them – elevating them above the plane of the banal, the uninspired, the undaring. In many ways, the whole typology is only a few baby steps along the giant path it needs to tread until tall buildings really do deliver their full potential, not only for cities but for humanity as a whole; in urbanistic and social, as well as energy terms. Against that backdrop, however, many of the projects profiled in this book do point the way forward for potentially improving tall buildings in some way. None of them may yet be the singular, all-singing, all-dancing answer, but each should be heralded for the special inventiveness they embrace.

This can certainly be said of all of the Finalists – projects the jury considers worthy of "winner" status – embraced in this book, a few of which I will highlight here.

In the Americas category, the Tree House Residence Hall in Boston is simply a delight aesthetically. A residential building for art students, inspired by Gustav Klimt's *Tree of Life*, the building goes way beyond this conceptual metaphor to somehow encapsulate in its physical expression the envisaged life and inhabitants of the program within. The façade expression itself speaks of fun, youth, and education. Learning that the program embraces communal spaces such as a "Pajama Floor" only adds to that impression.

In Asia, the PARKROYAL on Pickering in Singapore is the latest expression of tangible sustainable design

Left: Asia & Australasia Finalist Pearl River Tower, Guangzhou. The building's form was designed to maximize wind flow through recesses housing wind turbines.

Opposite Top: The inaugural 10 Year Award winner 30 St Mary Axe, London. Not only now a strong symbol for London, the building has paved the way for the acceptance of tall buildings within Europe's historic cities.

Opposite Bottom: Innovation Award co-winner, Broad Sustainable Building Prefabricated Construction Method. This construction method allowed for the construction of a 30-story hotel in just 15 days on-site.

principles and vertical greenery that is becoming increasingly evident in both Singapore and in the work of WOHA, which I believe is one of the best architectural practices working in tall buildings today. The use of tropical greenery within the building envelope and in its external expression, which we have come to associate strongly with WOHA's work, is a quite remarkable element in this building – 215% of the site area is given to green areas at height. Inspired by terraced rice paddies, the podium – containing many of the hotel facilities – is sculpted into a "terraformed" inverted landscape that extrudes in and out of the building, gives some stunning ground-floor vistas, and creates a building which feels, quite literally, at one with nature.

Pearl River Tower in Guangzhou also strives for sustainability like the aforementioned example, but does so through the embrace of technology – specifically through the manipulation of the building form and skin to channel wind energy into its turbine-filled recesses at levels 25 and 51. The completion of Pearl River Tower marks an important moment historically, since it was originally intended as the world's first "net-zero energy" skyscraper. The reality, as explained quite candidly by the architects, falls somewhat short of that ideal. Due to a number of unforeseen factors, political and otherwise, sustainable approaches are much easier to embrace at the design stage than to implement in reality. Pearl River is a fantastic building on many levels, but it also serves to remind us that, despite the oft-used rhetoric from the industry, we are a long way from delivering net-zero or net-positive energy tall buildings.

The reality is that we might never achieve that environmental "Holy Grail" in tall buildings, and almost certainly we will likely not achieve it if the embodied energy of the materials to construct the building – and other aspects of the whole building life cycle – are factored into the equation. The best we can perhaps do is strive to reduce both embodied and operating energy to the absolute minimums required, and then take solace in the fact that it is the bigger urban-scale contributors outside the building – the benefits of increased urban density, reduced infrastructure, shared facilities, better land use – that is tipping the scales in favor of the tall building, rather than the building acting alone. It is certainly this area – the building's full integration into the infrastructure of the city – where we need to increase our efforts.

Partly because awards are usually based predominantly on design *intent*, and the confusing rhetoric that sometimes comes from the industry on sustainability, this year we have created the new CTBUH "10 Year Award," which looks to award proven *performance* over at least a decade of building operation, rather than just design intent. That performance can be demonstrated against one or more of a wide number of fields, such as reduced operating energy consumption, contribution to the urban realm, embrace of the project as a piece of cultural iconography, etc. We are lucky that the inaugural winner of the award – the 30 St Mary Axe, or "Gherkin" building in London – exhibits all of those achievements. Beyond this, the Gherkin, more than perhaps any other modern tall building, paved the way for the possible acceptance of tall buildings within historic cities in Europe, by the public as well as the planning authorities. It positively paved the way for a whole raft of tall buildings in and beyond London. It was not so long ago that the under-construction Gherkin was criticized and derided by the seeming majority. Within a decade, however, it has been very positively embraced into the public conscience, to the point that some believe it is now worthy of listed status and view protection, not unlike the nearby St. Paul's Cathedral that has shaped much of the high-rise fabric of London. Quite a turn-around, indeed.

If the 10 Year Award is a new award, then the Innovation Award, only in its second year, already seems well established, if the number and quality of submissions for this category in 2013 are any indication. It seems the industry is brimming with innovative ideas and implementations across almost every field. For the first time in a CTBUH awards category ever, the jury this year simply could not determine which of two submissions was more worthy of the title of Innovation Winner, and thus decided to co-award two innovations, both of which could potentially revolutionize the industry: the Broad Sustainable Building Prefabricated Construction Method, and KONE's UltraRope system.

There are few moments in the tall building world in recent years as jaw-droppingly memorable as the day we all opened the YouTube video link to watch China BROAD Group's 30-story T30 Hotel being constructed on-site in just 15 days. Anyone with an interest in tall

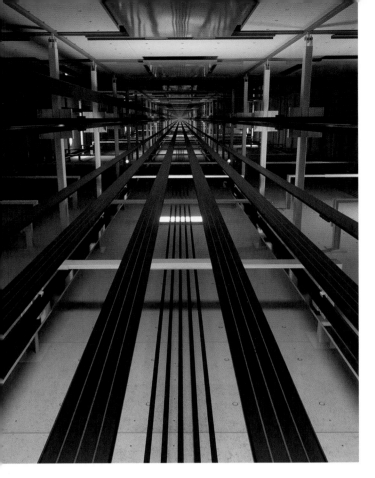

Left: Innovation Award co-winner, KONE UltraRope. The carbon-fiber ropes allow for an elevator's single-shaft height to reach 1000+ meters, effectively doubling the maximum shaft height within tall buildings.

building construction – even with a prerequisite dose of skepticism thrown in – could see that this was an unprecedented achievement on an international scale. Since then, the BROAD Group has employed the system on numerous buildings, even planning to erect the "world's next tallest" (at 838 meters) on-site in just nine months. Whether that latter plan ever does become a reality, or whether the myriad wider implications the proposal raises can be addressed, remains to be seen, but that is not to detract from the considerable achievements of the construction methodology so far.

Having visited both the fabrication plant and the T30 building, I can confirm that these are very real – and high-quality – endeavors. The act of mass-producing identical, prefabricated tall buildings throughout our cities raises numerous valid concerns – the further cultural homogenization of already rapidly homogenizing cities chief among them – but it is beyond doubt that the lessons from BROAD Group's system of on-site construction can benefit all tall buildings.

KONE's carbon-fiber UltraRope was launched earlier this year with a good deal of fanfare around its breakthrough achievement: a 1,000-meter-tall building can now be serviced through a single elevator rise. Currently, the limit of technology with steel rope – due to the size and weight of the rope itself – is about 500 meters. Though that benefit clearly has its eye to the future (with the 1,000 meter+ Kingdom Tower in Jeddah now starting on-site), it was the efficiencies and energy-saving potential for all tall buildings that more strongly influenced the jury to recognize this product as innovative. Implementable electromagnetic and other propulsion technologies may someday evolve to supplant the somewhat archaic, almost two-century-old notion of pulling an elevator car by a rope. Until then, the KONE UltraRope is possibly one of the biggest breakthroughs in the elevator world since Otis originally demonstrated the Safety Elevator.

And that is what the CTBUH Innovation Award is really about – recognizing those projects that are truly innovative, and have the chance to revolutionize parts of our industry. On that note, when it comes to both innovation and revolution, it is hard to argue against the two winners of our lifetime achievement awards this year; Clyde N. Baker, Jr. of AECOM STS Engineering for the Fazlur R. Khan Medal, and Henry N. Cobb of Pei Cobb Freed & Partners for the Lynn S. Beedle Award.

It is hard to think of a geotechnical engineer more accomplished than Clyde Baker. Responsible for the foundation design of six of the tallest 12 buildings in the world, he has been a constant source of professionalism and inspiration for the geotechnical industry for over half a century. And, in a similar vein, Henry Cobb – while perhaps the lesser known partner to the I.M. Pei side of the practice – has gathered respect for his accomplishments throughout the tall building world.

All CTBUH Trustees, who determine the lifetime achievement awards, listened with fascination as Trustee William Baker of Skidmore Owings & Merrill recounted his early experiences of "Harry" Cobb during the Boston Hancock Place fiasco in the early 1970s. In a world of increasing litigation and finger-pointing, Harry stood up for the right way forward on that project – at great risk to himself – simply because it was the right thing to do. This earned him accolades in the architectural press. William LeMessurier, a structural engineer who worked on the Hancock project with Cobb, told *Architecture* magazine in 1988 that Henry was "not only responsible, but inspiring. …Harry Cobb was an absolute model of how a professional should behave in this kind of situation."

Tall Buildings and Population Density: A Snapshot of Regional Comparisons

Each year in this awards book introduction, we move beyond the awards themselves, to put their achievements in the context of wider trends happening globally. This year we have undertaken research on the four regions represented by the Best Tall Buildings recipients – Canada, China, Europe, and the Middle East. As you will see from the study represented over the following pages and in the summary table (see below), with the exception of the somewhat-smaller Middle East, the other three regions are all roughly the same in geographical area, but each is markedly different in both its population density and embrace of tall buildings. The comparisons are interesting because, although the world is generally becoming more urbanized and cities growing more vertical, both population and land use (two of the main drivers for tall buildings) across the regions are radically different.

If we begin with a consideration of total population and population density then China has approximately

Country/Region Comparisons[1]

	Demographics			Cities[2]			150m+ Buildings[3] by 2015			World Percentage		
Country/Region	Total Land Area (km²)[4]	Total Population[5]	Density (per km²)	1,000,000+ Population	At Least One 150m+ Building by 2015	Most 150m+ Buildings by 2015	Total Number	Average Height	Tallest	Of Total Population	Of Land Area	Of 150m+ Buildings by 2015
Canada	9,094,037	33,476,688	3.7	6	7	Toronto (46) Calgary (16)	75	187.9 m	First Bank Tower (298 m, 1975)	0.5%	6.1%	2.2%
China	9,570,983	1,339,724,852	140.0	171	69	Hong Kong (293) Shanghai (127)	1,264	199.3 m	Shanghai Tower (632 m, 2014)	19.0%	6.4%	36.8%
Europe	9,799,870	720,006,668	73.5	59	26	Istanbul (36) Moscow (36)	165	190.6 m	Vostok Tower (360 m, 2014)	10.1%	6.6%	4.8%
Middle East	7,119,839	381,402,626	53.6	38	22	Dubai (151) Abu Dhabi (32)	295	362.2 m	Burj Khalifa (828 m, 2010)	5.4%	4.8%	8.6%
United States (for reference)	9,158,960	316,668,567	34.6	63[6]	54	New York City (237) Chicago (115)	685	181.2 m	One World Trade Center (541 m, 2014)[8]	4.5%	6.2%	20.0%
World	148,940,000	7,095,217,980	47.7	503[7]	289	Hong Kong (293) New York City (237)	3,434	197.0 m	Burj Khalifa (828 m, 2010)	–	–	–

Above: Table showing comparisons of population, area, and 150 meter+ buildings by 2015 for the four countries/regions represented by this year's Best Tall Buildings award winners, showing also the United States and World for comparison. Detailed studies of each country/region can be found on the following pages.

1. For sources of all data for country/regional figures, please refer to the individual regional studies on the following pages.
2. "City" refers to the Urban Agglomeration, generally defined as a central city and neighboring communities linked to it by continuous built-up areas or commuters.
3. The focus on buildings over 150 meters is driven by the need to ensure accuracy of data, rather than suggesting that this is the threshold for a tall building. All tall building data is taken from the CTBUH Skyscraper Center as of July 26th, 2013.
4. Unless otherwise noted, all land area data is taken from the UN Demographic Yearbook 2009-2010.
5. Unless otherwise noted, all population data is taken from the CIA World Factbook.
6. Number of 1,000,000+ cities for the United States is based on data from the 2010 United States Census for Primary Statistical Areas, defined as metropolitan areas that are not components of any more extensive defined metropolitan areas.
7. Estimate of 1,000,000+ cities for the World is based on data from Thomas Brinkhoff: Major Agglomerations of the World, at http://www.citypopulation.de.
8. The height of One World Trade Center has not yet been ratified by the CTBUH Height Committee.

40 times more people in roughly the same land area as Canada (1.34 billion people compared with Canada's 33.5 million). This equates to a population density for China of 140 people per square kilometer, compared with Canada's 3.7 people per square kilometer. Europe is approximately half as densely populated as China, with a population density of 74 people per square kilometer, while the Middle East region has a population density of 54 people per square kilometer.

If we move to the next scale – that of cities – then the differences are even more marked. It should be noted that, in considering cities, we have considered the "Urban Agglomeration" rather than the more arbitrary local definition of what constitutes a particular city. An urban agglomeration is defined as an extended city or town area comprising the built-up area of the central municipality as well as any cities and suburbs linked in a continuous urban area.[1]

Canada thus has only six urban agglomerations greater than 1 million inhabitants, the Middle East has 38, Europe has 59, and China alone has an incredible 171 cities over 1 million people – a number set to grow significantly over the coming years. To bring in another comparison, the USA has 63 cities greater than 1 million inhabitants. Of course these comparisons are not fully equal; Canada has the approximate same land area as China but, of course, huge parts of the country are inhospitable to inhabitation because of the extremely cold climate. However, both countries are developing their cities vertically, so the comparisons are interesting.

When it comes to "skyscraper cities" – those defined for this study as any city with at least one building greater than 150 meters in height – then a similar pattern plays out. By 2015 it is projected that Canada will have seven cities with tall buildings greater than 150 meters, the Middle East will have 22, Europe will have 26, and China will have 69. In terms of the actual buildings themselves, China is projected to have a staggering 1,264 buildings over 150 meters in height, some 17 times greater than Canada will have with 75 such buildings, eight times greater than the whole of Europe with 165, and four times greater than the Middle East region, with 295. China's buildings over 150 meters will actually constitute 37% of the global total, which is 3,434.

In pure height terms, the Middle East wins out in both the height of the tallest building (the Burj Khalifa at 828 meters) and the average height of all buildings over 150 meters by 2015 – 362 meters. The other regions, including China, will have approximately half this average height number, suggesting that great height may be more of a driving objective in the Middle East region that other parts of the world. We will now look in more detail within the regions themselves.

Canada
Since 2005, Canada has been in the midst of a major tall building boom. Twenty-nine buildings taller than 150 meters have been built since then, with an additional 18 planned to be completed by 2015. In 2012 alone, Canada added four buildings taller than 200 meters, which is the most it has ever completed in a single year. In contrast, the US completed only two buildings over 200 meters in 2012.

The epicenter of Canadian tall building development is Toronto, where 17 buildings taller than 150 meters are under construction, more than any other city in the western hemisphere. Toronto is projected to have 46 buildings taller than 150 meters by 2015, up from 13 buildings in 2005. Currently, Toronto accounts for 52.6 percent of the buildings taller than 150 meters in Canada, while comprising only 14.6 percent of the population. This share of tall buildings is expected to rise to 61.3 percent by the end of 2015.

The development in Canada also reflects a global shift in the fundamental role of tall buildings around the world; a shift away from pure office buildings towards

---

[1] The reader should note that, in some cases where there is a distinct city within an urban agglomeration that contains its own skyscrapers, then this is pulled out as its own urban entity. For example, Mississauga is analyzed separately to Toronto, though it certainly falls within the Toronto urban agglomeration, and similar has been done with Burnaby/Vancouver and Shenzhen/Guangzhou/Hong Kong.

Canada: 150m+ Buildings by 2015

1. The focus on buildings over 150 meters is driven by the need to ensure accuracy of data, rather than suggesting that this is the threshold for a tall building.
2. All tall building data is taken from the CTBUH Skyscraper Center as of July 26th, 2013.
3. Unless otherwise noted, all population data is taken from the Canada Census, 2011.
4. Unless otherwise noted, all land area data is taken from the UN Demographic Yearbook 2009–2010.
5. "City" refers to the Urban Agglomeration, defined as an extended city or town area comprising the built-up area of the central place (usually a municipality) plus any cities and suburbs linked in a continuous urban area. Estimate of 1,000,000+ cities is based on data from the Canada Census, 2011.
6. Since Mississauga and Burnaby have their own 150m+ buildings, they have been shown separately in this study from the Urban Agglomerations which contain them.

Key
- X Buildings / X.X% of total: number of **buildings** 150m+ in height in city and **percentage** of country total by the year 2015
- total **population** of city and **percentage** of country population
- **height** of city's tallest building by the year 2015
- **silhouette** of city's tallest building by the year 2015
- **average height** of all buildings 150m+ in city by the year 2015

Canada Totals
Total Population:[3] **33,476,688**
Total Land Area:[4] **9,094,037 km²**
Population Density: **3.7 people/km²**
Cities of 1,000,000+ Population:[5] **6**

Est. by 2015…
Cities with at least one 150m+ building: **7**
City with the most 150m+ buildings: **Toronto (46)**
Total 150m+ buildings: **75**
Tallest building height: **298 m**
Average height of 150m+ buildings: **188 m**

Vancouver — 2,090,110 / 6.2% of pop. / 2 Buildings / 2.7% of total
Burnaby[6] — 223,218 / 0.7% of pop. / 1 Building / 1.3% of total
Calgary — 1,214,839 / 3.6% of pop. / 16 Buildings / 21.3% of total
Toronto — 4,869,621 / 14.6% of pop. / 46 Buildings / 61.3% of total
Montreal — 3,824,221 / 11.4% of pop. / 7 Buildings / 9.3% of total
Mississauga[6] — 713,443 / 2.1% of pop. / 2 Buildings / 2.7% of total
St. Catharines-Niagara Falls — 392,184 / 1.2% of pop. / 1 Building / 1.3% of total
Rest of Canada — 20,149,052 / 60.2% of pop. / 0 Buildings / 0.0% of total

First Bank Tower 1975 Toronto	The Bow 2012 Calgary	Le 1250 Boulevard Rene-Levesque, 1992 Montreal	Shangri-La Vancouver 2009 Vancouver	Niagara Falls Hilton 2009 St. Catharines-Niagara Falls	Absolute World Building D 2012 Mississauga	Sovereign 2014 Burnaby
298 m / 189 m	237 m / 182 m	226 m / 186 m	201 m / 180 m	177 m	176 m / 167 m	156 m

China: 150m+ Buildings by 2015

The cities shown in the diagram are the top 15 cities by number of buildings taller than 150 meters that are expected to be completed by the end of 2015. The remaining 54 cities with at least one such building are listed as a group.

1. The focus on buildings over 150 meters is driven by the need to ensure accuracy of data, rather than suggesting that this is the threshold for a tall building.
2. All tall building data is taken from the CTBUH Skyscraper Center as of July 26th, 2013.
3. Unless otherwise noted, all population data is taken from the 2010 Chinese Census by the National Bureau of Statistics of China.
4. Unless otherwise noted, all land area data is taken from the UN Demographic Yearbook 2009–2010.
5. "City" refers to the Urban Area as defined by the National Bureau of Statistics of China, which is akin to the Urban Agglomeration criteria used throughout this study. Number of 1,000,000+ cities is an estimate by People's Daily.

China Totals

Total Population:[3] **1,339,724,852**
Total Land Area:[4] **9,570,983 km²**
Population Density: **140.0 people/km²**
Cities of 1,000,000+ Population:[5] **171**

Est. by 2015…
Cities with at least one 150m+ building: **69**
City with the most 150m+ buildings: **Hong Kong (293)**
Total 150m+ buildings: **1,264**
Tallest building height: **632 m**
Average height of 150m+ buildings: **199 m**

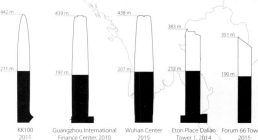

residential and mixed-use. In 2001, 26 of the 27 buildings taller than 150 meters in Canada were offices or hotels. As of July 2013, 17 of the 18 buildings taller than 150 meters, currently under construction, are entirely or partially residential. In addition, all 16 of the towers taller than 150 meters under construction in Toronto are entirely or partially residential.

What accounts for these tall-building trends in Canada? The positive tall-residential trend, already well under way in Vancouver, where the building culture is highly reflective of the substantial Chinese population in that city, appears to have spread to Toronto. As in many other North American cities, a new generation is becoming interested in the benefits of walkable urbanism in Toronto. Formerly low-density suburban areas are building up around transit hubs, and formerly shunned inner-city areas are being rejuvenated.

Calgary rides a double-edged sword that points skyward. The city has highly progressive environmental and planning regulations and a well-used light-rail system, which support dense urbanity. Yet much of the new office growth in Calgary is driven by the resurgent energy business, as the oil and gas industries push ever-northward in search of the earth's remaining fossil-fuel deposits. These companies have strong reasons to grow and consolidate their offices in towers that reflect their economic prominence, but they also seem to feel an obligation to create high-performing buildings with good civic components.

On the subject of tall building design quality, it is interesting to note that, in addition to The Bow winning the Best Tall Building Americas award this year, two other Canadian buildings have won the award in recent years – Absolute Towers in Mississauga (2012) and Manitoba Hydro Place in Winnipeg (2009). Canada has thus won the Americas award in three of the past five years.

China

With over 1.3 billion citizens and a rapidly urbanizing population, China is developing tall buildings in greater volume and with greater speed than any other country globally. Currently it has 912 buildings over 150 meters in height, with 290 buildings currently under construction and expected to be completed by the end of 2015. In comparison, the entire rest of the world has only 356 buildings over 150 meters expected to complete by that time. Thus, by the end of 2015, it is estimated that one in three buildings taller than 150 meters in the world will be located in China.

The distribution of tall buildings across the cities of China is wide. Though Hong Kong, Shanghai, and Guangzhou are the three cities containing the most tall buildings currently, these are not the only epicenters of development. By the end of 2015, at least 69 Chinese cities will contain one building over 150 meters in height. Twenty-three of those 69 cities will have at least 10 buildings taller than 150 meters.

This correlates with two major trends currently at play in China. The first is that the government has a policy to move 250 million rural residents into cities by 2025, with the objective of shifting China from an agrarian economy to an industrial/consumer economy. The second is that Chinese manufacturing is shifting away from the major cities on the coasts and into the interior, in search of lower labor costs. The confluence of these trends would seem to point toward increased urbanization and tall-building activity in regional cities, the names of many of which are unfamiliar to Western ears today, but may not be tomorrow. One such city is Shenyang, which is expected to have 81 buildings over 150 meters tall by 2015, ranking it fifth in the country with 6.4 percent.

Hong Kong, with 31.9 percent of all of China's buildings taller than 150 meters, and Shanghai, with 12.7 percent, still dominate. However, this dominance has been diminishing over recent years. By the end of 2015, these percentages are predicted to shrink to 23.2 percent and 10.1 percent, as other cities in China grow and develop.

In terms of tall building design quality, in addition to the CCTV Building winning the Best Tall Building award for Asia & Australasia this year, three other Chinese buildings have won the award in recent years – Guangzhou International Finance Center in Guangzhou (2011), Linked Hybrid Bridge in Beijing (2009), and Shanghai World Financial Center in Shanghai (2008). Thus China has won the Asia & Australasia award four times out of the past six years.

Europe
Comprising all or part of 47 countries, Europe is the largest region in this study by land area, and the second largest by population, with 720 million people, or 10.1 percent of the world total. However, the region as a whole is projected to represent only 4.8 percent of buildings taller than 150 meters by 2015. This figure is only a little more than twice the 2.2 percent that will be held by Canada, a region with less than one-twentieth of Europe's population.

Despite this low tall-building share, Europe has experienced a fairly consistent increase in construction of buildings over 150 meters since the late 1980s. In 1995, the number of buildings 150 meters or taller in Europe was 32. By 2005, this number had more than doubled to 65, and by 2015 it is expected to more than double again to 165. There are currently 55 buildings over 150 meters under construction with expected completion by 2015, the largest number in the region's history.

The lion's share of this tall building development is characterized in two countries and, specifically, two cities – Istanbul, Turkey and Moscow, Russia. By the year 2015, these two cities are predicted to retain a collective 44 percent of all buildings taller than 150 meters in Europe. In comparison, their collective population share is merely 16.7 percent of the region's total. Istanbul, in particular, has 18 buildings taller than 150 meters that are under construction with expected completion by 2015, the largest building projection for any city outside of Asia or the Middle East.

Europe is still strong in constructing office buildings, with 38 of the 55 buildings 150 meters+ under construction with expected completion by 2015, containing at least some planned office space in their program. Residential tall building construction has also been increasing, with eight of the 10 tallest buildings by 2015 projected to contain at least some residential element in their program. This may be indicative of a warming trend toward residential high-rises, which in postwar Europe had generally held negative associations of social housing, urban strife, and Brutalist architectural styles, such that now they are often seen as highly desirable places to live.

There is a lack of significant correlation between population density and tall building distribution in Europe. The 15 countries projected to have at least one building taller than 150 meters by 2015 have an overall average density of 75.7 people per square kilometer; while the remaining 32 countries in Europe are just shy of 66.1 people per square kilometer. Interestingly, Belgium has a density of 345 people per square kilometer, which is more than all three of the other surveyed regions, but is not projected to lay claim to any building taller than 150 meters by 2015.

As a whole, then, Europe is showing a much increased appetite for building tall. On the one hand, globalization has wrought the same changes in Europe as elsewhere – the standard Class-A office space requirements have resulted in agglomerations such as La Défense, Canary Wharf, and Moscow-City. But there is unquestionably a European brand of skyscraper that has emerged more recently. This archetype has graduated from square-plan, generic towers in urban-peripheral "skyscraper districts," to highly contextual, sculpted towers ever closer to city centers, bending and twisting to accommodate irregular plot shapes and view corridors to landmarks in centuries-old metropolises.

The Shard marks the fifth time the UK has won the CTBUH Best Tall Building Award for Europe in the past seven years, with the other winners being Broadcasting Place in Leeds (2010), The Broadgate Tower in London (2009), 51 Lime Street in London (2008), and the Beetham Hilton Tower in Manchester (2007).

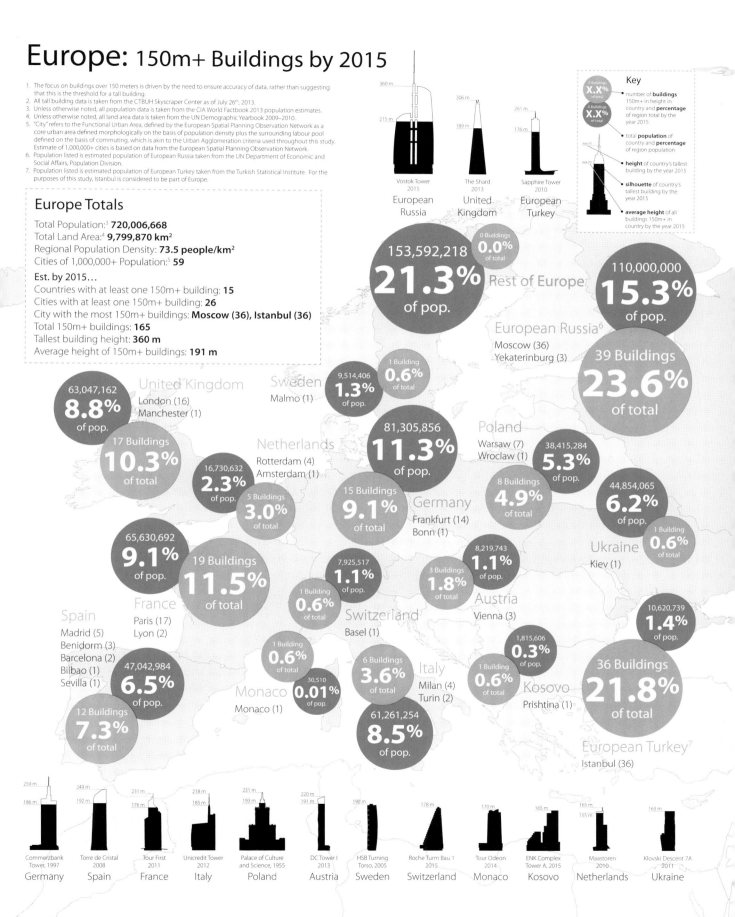

Middle East: 150m+ Buildings by 2015

1. The focus on buildings over 150 meters is driven by the need to ensure accuracy of data, rather than suggesting that this is the threshold for a tall building.
2. All tall building data is taken from the CTBUH Skyscraper Center as of July 26th, 2013.
3. Unless otherwise noted, all population data is taken from the CIA World Factbook 2013 population estimates.
4. Unless otherwise noted, all land area data is taken from the UN Demographic Yearbook 2009–2010.
5. "City" refers to the Urban Agglomeration, defined as an extended city or town area comprising the built-up area of the central place (usually a municipality) plus any cities and suburbs linked in a continuous urban area. Estimate of 1,000,000+ cities is based on data from the World Gazetteer.
6. Population listed is estimated population of Asiatic Turkey taken from the Turkish Statistical Institute. For the purposes of this study, Istanbul is considered to be part of Europe.

Key
- X.X% of total: number of **buildings** 150m+ in height in country and **percentage** of region total by the year 2015
- X.X% of total: total **population** of country and **percentage** of region population
- **height** of country's tallest building by the year 2015
- **silhouette** of country's tallest building by the year 2015
- **average height** of all buildings 150m+ in country by the year 2015

Middle East Totals
- Total Population:³ **381,402,626**
- Total Land Area:⁴ **7,119,839 km²**
- Regional Population Density: **53.6 people/km²**
- Cities of 1,000,000+ Population:⁵ **38**

Est. by 2015…
- Countries with at least one 150m+ building: **10**
- Cities with at least one 150m+ building: **22**
- City with the most 150m+ buildings: **Dubai (151)**
- Total 150m+ buildings: **295**
- Tallest building height: **828 m**
- Average height of 150m+ buildings: **218 m**

Asiatic Turkey⁶
- Ankara (3)
- Izmir (2)
- Konya (1)
- Mersin (1)

70,073,746 — 18.4% of pop.
7 Buildings — 2.4% of total

Iran
- Kish (1)
- Tehran (1)

79,853,900 — 20.9% of pop.
2 Buildings — 0.7% of total

Lebanon
- Beirut (1)

4,131,583 — 1.1% of pop.
1 Building — 0.3% of total

Jordan
- Amman (1)

6,482,081 — 1.7% of pop.
1 Building — 0.3% of total

Israel
- Tel Aviv (6)
- Ramat Gan (2)
- Bnei Brak (1)

7,707,042 — 2.0% of pop.
9 Buildings — 3.1% of total

Kuwait
- Kuwait City (14)

4,695,316 — 1.2% of pop.
14 Buildings — 4.8% of total

Bahrain
- Manama (11)

1,281,332 — 0.3% of pop.
11 Buildings — 3.7% of total

Rest of Middle East
172,721,627 — 45.3% of pop.
0 Buildings — 0.0% of total

Saudi Arabia
- Riyadh (11)
- Jeddah (9)
- Mecca (7)
- Al Khobar (1)

26,939,583 — 7.1% of pop.
28 Buildings — 9.5% of total

Qatar
- Doha (27)

2,042,444 — 0.5% of pop.
27 Buildings — 9.2% of total

United Arab Emirates
- Dubai (151)
- Abu Dhabi (32)
- Sharjah (9)
- Al Fujayrah (2)
- Ajman (1)

5,473,972 — 1.4% of pop.
195 Buildings — 66.1% of total

Tallest Buildings by Country

Building	Height	Year	Country
Burj Khalifa	828 m (avg 217 m)	2010	United Arab Emirates
Makkah Royal Clock Tower Hotel	601 m (avg 267 m)	2012	Saudi Arabia
Al Hamra Tower	413 m (avg 216 m)	2011	Kuwait
Aspire Tower	300 m (avg 208 m)	2007	Qatar
Four Seasons Hotel	270 m (avg 215 m)	2014	Bahrain
City Gate Tower	244 m (avg 172 m)	2001	Israel
Flower of the East Hotel	210 m (avg 186 m)	2014	Iran
Folkart Tower A	200 m (avg 172 m)	2014	Asiatic Turkey
Amman Rotana Hotel	188 m	2013	Jordan
Sama Beirut	187 m	2014	Lebanon

Middle East

With 381 million people, the Middle East represents only 5.4 percent of the world's population and has the smallest land area of any region in this study. However, the area has been a major center of tall building development in recent times. In 2005, the region only had 33 buildings taller than 150 meters, or 2.1 percent of the world total. By 2015, this number is expected to increase to 295 buildings, which will give it 8.6 percent of the world total. This percentage increase is nearly seven times greater than that which has occurred in China in the same amount of time. The region's boom is especially focused on supertall buildings, of which it will have added 32 buildings between 2005 and 2015, and will be home to approximately 26% of all supertalls by that time.

By far the main epicenter of tall buildings in the region is the United Arab Emirates, which is projected to hold 66.1 percent of all buildings 150 meters or taller in the region by 2015. No other country will contain more than 10 percent by this time. This concentration should decrease in the future as 33 of the 60 under construction buildings over 150 meters tall, with expected completion by 2015, will be located in countries outside the UAE experiencing their own tall building booms. Saudi Arabia, in particular, currently only has 13 buildings over 150 meters, but it is expected to more than double this number by completing another 15 by 2015.

Unlike the other surveyed regions, overall country population density is not a good indicator of tall building distribution in the Middle East. The 10 countries in the Middle East with at least one building taller than 150 meters by 2015 have an average density of 36.9 people per square kilometer, while the rest of the region has an average density of 118 people per square kilometer. Interestingly, Egypt, which is the largest country in the region by both population and area, is projected to have no buildings taller than 150 meters by the end of 2015.

The Middle East, like Canada, is following the trend towards incorporating more residential-based tall buildings, with 70.3 percent of all buildings over 150 meters by 2015 having some residential element included within its program. When only considering supertall buildings, 300 meters or higher, the residential-based building percentage is even higher, at 77.8 percent.

What can we draw from this? Clearly, the drivers of tall in the Middle East are not directly correlated to mass urbanization as in China. Given the prevalence of extremely tall buildings with significant "vanity heights" (non-occupiable areas exclusively devoted to gaining altitude), one might perhaps conclude that sheer height and architectural distinction are being used to establish global symbols of wealth and power for individuals, companies, cities, and countries in the region.

In terms of the CTBUH Best Tall Building award for the Middle East & Africa, the recipient cities have been somewhat distributed across the Middle East region (with no winners in Africa so far); in addition to Sowwah Square winning this year for Abu Dhabi, other recent winners have included Doha in 2012 and 2009 (Doha Tower and Tornado Tower respectively), Dubai in 2011 and 2010 (The Index and Burj Khalifa respectively), and Manama in 2008 (Bahrain World Trade Center).

Conclusion

Clearly then, tall buildings are becoming more prevalent in more corners of the world. However, despite the inventiveness in many projects contained in this publication, the fact remains that many tall buildings are not brilliant pieces of architecture, and are generally homogenizing the urban centers they populate. We now have the ability to go much higher, with much greater speed than ever before, but we must also build better – by directly relating every building to its physical, cultural, and environmental context. The design quality and innovation represented by this year's award winners set the bar so much higher for what a tall building can be, and give me confidence that we will continue to develop the typology better into the future.

CTBUH Best Tall Building
Awards Criteria

The Council on Tall Buildings and Urban Habitat initiated its Awards Program in 2001, with the creation of the Lynn S. Beedle Lifetime Achievement Award. It began recognizing the team achievement in tall building projects by issuing Best Tall Building Awards in 2007, to give recognition to projects that have made extraordinary contributions to the advancement of tall buildings and the urban environment, and that achieve sustainability at the highest and broadest level. It issues four regional awards each year, and from the "regional" awards, one project is awarded the honor of overall "Best Tall Building Worldwide," which is announced at the awards ceremony.

The winning projects must exhibit processes and/or innovations that have added to the profession of design and enhance cities and the lives of their inhabitants. Some of the criteria for submission are outlined below. It is important to note that, with the exception of the first point (regarding completion date eligibility), a project does not necessarily need to meet every listed criteria. Submissions should demonstrate strengths in areas that are applicable:

1) The project must be completed (topped out architecturally, fully clad, and at least occupiable) no earlier than the 1st of January of the previous year, and no later than the 1st of October of the current awards year (e.g., January 1, 2012 and October 1, 2013 for the 2013 awards).

2) The project advances seamless integration of architectural form, structure, building systems, sustainable design strategies, and life safety for its occupants.

3) The project exhibits sustainable qualities at a broad level:
Environment: Minimize effects on the natural environment through proper site utilization, innovative uses of materials, energy reduction, use of alternative energy sources, reduced emissions, and water consumption.
People: Must have a positive effect on the inhabitants and the quality of human life.
Community: Must demonstrate relevance to the contemporary and future needs of the community in which it is located.
Economic: The building should add economic vitality to its occupants, owner, and community.

4) The project must achieve a high standard of excellence and quality in its realization.

5) The site planning and response to its immediate context must ensure rich and meaningful urban environments.

6) The contributions of the project should be generally consistent with the values and mission of the CTBUH.

Note: Awards in some categories may not be conferred on an annual basis if the criteria cannot be clearly met or demonstrated through the submittal.

Best Tall Building
Americas

Winner
Best Tall Building Americas

The Bow
Calgary, Canada

Completion Date: 2012
Height: 237 m (779 ft)
Stories: 57
Area: 199,781 sq m (2,150,420 sq ft)
Use: Office
Owner: H + R Reit
Developer: Matthews Southwest
Architect: Foster + Partners (design); Zeidler Partnership Architects (architect of record)
Structural Engineer: Yolles
MEP Engineer: Cosentini Associates
Main Contractor: Ledcor Construction
Other Consultants: Altus Group (cost); Brook Van Dalen (façade); Carson McCulloch (landscape); Cerami Associates (acoustics); Claude Engle Lighting Design (lighting); Gensler (interiors); Kellam Berg (civil); KJA (vertical transportation); Leber Rubes (fire); RWDI (wind); Transsolar (energy concept)

> *"The Bow's passive approach to solar control and ventilation are implicit in its form, supported by an interesting structural system that is legible on the building's exterior."*
>
> Jeanne Gang, Jury Chair, Studio Gang Architects

The Bow is the first phase of a mixed-use master plan for the regeneration of two entire city blocks on the east side of Centre Street, a major axis through downtown Calgary. Providing a headquarters for a major energy company, its form was shaped by both environmental and organizational analysis. The tower faces south, curving toward the sun to take advantage of daylight and heat, while the resulting bow-shaped plan that gives the tower its name maximizes the perimeter for cellular offices with views of the Rocky Mountains.

The aerodynamic crescent shape significantly reduces exterior wind resistance, downdrafts, and urban wind tunnels to create a comfortable public plaza at the tower's base. Thus, the arc-shaped form helps to define this large civic space; the south-facing plaza will create a popular public space for use all year round.

At 237 meters, The Bow is the tallest tower in Calgary, but it is equally significant in terms of the lateral connections it establishes with the surrounding buildings

Previous Spread

Left: View of tower from southwest

Right: Interior view of the sky garden at level 54

Current Spread

Right: Section – the full-height atria connects the lobby and three sky gardens (highlighted in blue)

Opposite Top: View of tower in context

Opposite Bottom: Floor plan – sky garden at level 24

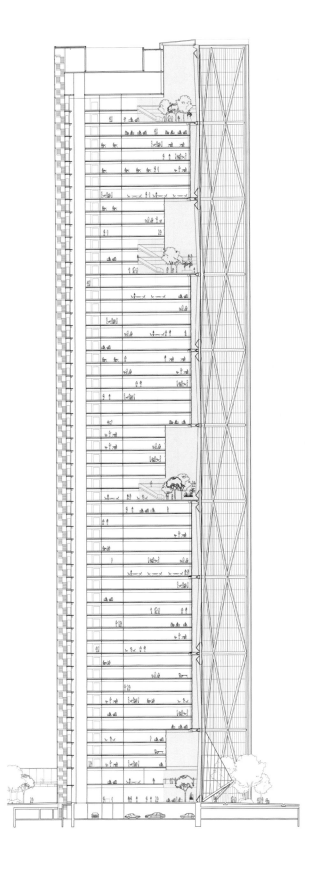

at its lower levels. Calgary is crisscrossed by a system of enclosed walkways which offers a retreat from the city's harsh winters. The tower is fused to these routes at three points. For example, the second floor is open to the public and integrates shops and cafés. Forming the only public connection over Centre Street, the scheme completes a vital link in the downtown pedestrian network.

Inside, the shape generates a floor plan that maximizes views and natural light, while providing a flexible, open workspace for its occupants. Where the building curves inwards, the glazed façade is pulled forward to create a series of atria that run the full height of the tower. Three sky gardens, which project into the atria at levels 24, 42, and 54, promote collaboration and bring a social dimension to the office spaces.

The gardens feature mature trees, seating, meeting rooms, and local lift cores – at each lobby, passengers travel to local groups of elevators, which serve all the floors within each "garden-level" building zone. This combination of elevator strategy and the incorporation of high-level green spaces encourages interaction and reasserts the social hubs that rise vertically through the building. At level 54, the building features a large 200-seat auditorium.

The atria provide an opportunity for several sustainable strategies that help reduce energy consumption. These spaces act as climatic buffer zones, insulating the building and helping to reduce energy consumption

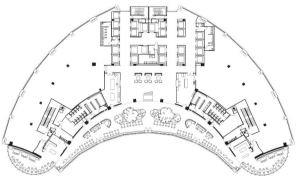

by approximately 30 percent. Excess heat from the office floors is channeled into the atria, while at the same time the sun's energy (given the atria's orientation) is harnessed. The atrium spaces act as a buffer zone between offices adjacent to the atrium and the exterior atrium glass wall, dramatically reducing energy consumption and the need for heating/cooling by exhausting heat upwards in summer and trapping heat in winter. Offices adjacent to the atrium have the ability to open windows into the atrium during the mild seasons.

Jury Statement

The Bow is both stunning as a form and functions well from an environmental and urban standpoint, especially in the context of a harsh northern climate. It serves as a rare example of an iconic design resulting from the most practical, yet creative, response to site constraints. The resolution of wind loading, light access, thermal comfort, and public space objectives has resulted in a solution that embodies synthesis but bears no hint of compromise.

A city known for dramatic weather changes is now graced with an elegant, glass-sheathed, year-round building that provides ample opportunities for occupants and the public to interface amid greenery and gracious views. The conventional design response to these conditions might have been to seal off the interior and adopt a "bunker-in-the-sky" mentality. But at The Bow, every design move does double or triple duty – the atrium, for example, is a living lung that also increases social serendipity – proving that a well-designed building can be efficient, beautiful, and generous all at once.

Left: Ground floor lobby
Opposite: View of tower from northeast

The orientation of the tower plays a critical role in the reduction of energy consumption. As the atrium façade of the towers faces south-southwest, the tower consumes 11 percent less energy for heating and cooling over the course of a year compared to towers with an atrium façade facing north. Even though the façade is oriented in the direction where the cooling requirement is highest, the solar energy received during the winter season compensates and actually reduces the overall annual energy requirement.

From a structural standpoint, this is the first time that a triangular diagrid has been applied to a curved skyscraper in North America. The structural system provides superior structural efficiency, while the diagonal and vertical steel frame reduces the overall weight of the steel, and thus the number and size of interior columns, while helping to break down the scale of the building visually.

"The overall design is largely successful due to its south-facing curve and centrally located communal spaces – both inside and outside the building – which minimize environmental impact, while maximizing community."

Karen Weigert, Juror, Chicago Chief Sustainability Officer

Finalist
Best Tall Building Americas

Devon Energy Center
Oklahoma City, United States of America

Completion Date: December 2012
Height: 257 m (844 ft)
Stories: 50
Area: 125,812 sq m (1,354,229 sq ft)
Use: Office
Owner: Devon Energy Corporation
Developer: Hines
Architect: Pickard Chilton (design); Kendall/Heaton Associates (architect of record)
Structural Engineer: Thornton Tomasetti
MEP Engineer: Cosentini Associates
Main Contractor: Holder Flintco Joint Venture
Other Consultants: Cerami Associates (acoustics); Gensler (interiors); Integrated Environmental Solutions, Ltd. (sustainability); Murase Associates (landscape); Persohn/Hahn Associates (vertical transportation); Professional Services Industries (geotechnical); The Office of James Burnett (landscape)

"It will surprise some that the quest for more sustainable high-rise buildings is being advanced in the heart of Oklahoma, but the Devon Energy Center is quietly doing just that."

Antony Wood, Juror, CTBUH

In 2006, Devon began planning its new headquarters on a 7.52-acre brownfield parking lot site on the southern edge of the Central Business District in Oklahoma City. The program included: a 50-story office tower; a podium comprising conference, training, and dining facilities; a 2.25-acre public park; a 300-seat publicly accessible auditorium; and a rotunda to serve as the complex's "town square." The entire ground level is open and accessible to the public.

Devon Energy Center creates a focal point for the company and Oklahoma City by integrating civic-scaled spaces as a vital component of its overall development. It consolidates Devon's Oklahoma City-based workforce into a single state-of-the-art facility with numerous amenities.

The three-sided tower evolved in part from Devon's desire to not "turn its back" on any part of the city. Its orientation and placement provide southern exposure to the park while minimizing solar gain. Its

Previous Spread
Left: View of tower from west
Right: Interior view of rotunda

Current Spread
Top Left: View of art wall within a multi-story atrium opening in the podium
Bottom Left: Typical tripartite floor plan
Opposite Top: 2.25 acre public park at the base of the tower
Opposite Bottom: Interior view of the main lobby

form resulted in highly efficient tripartite floor plates averaging 28,000 square feet that accommodate up to 12 full-corner offices. Responsive to the theme of a "right to light" for all occupants, the 10-foot floor-to-ceiling glazing allows daylight deep into the Tower as well as expansive views. All perimeter offices have floor-to-ceiling glass to maximize daylight. The curtain wall consists of continuous high-performance clear glass with a low-E coating that maximizes daylight, while also reducing heat gain.

With a highly articulated structure, the jewel-like Rotunda is a grand civic-scaled space with glass walls, a series of balconies, and sky-lit roof. It regularly serves as a venue for special events. Unifying the entire complex within the city, this Rotunda symbolically and literally connects the cardinal directions, punctuating the urban axis of Harvey Street. The Rotunda provides 12,522 square feet (1,163 square meters) of welcome assembly space and has hosted fundraisers, holiday balls, corporate dinners, and civic gatherings.

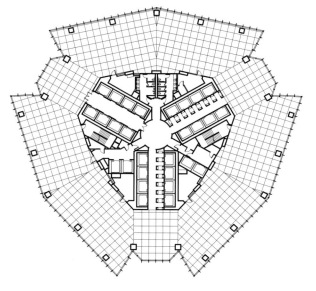

The podium contains training and meeting spaces as well as visitor and occupant services. A promenade extending its length creates a glazed day-lit interior corridor. At street level, it provides public access to various amenities, including restaurants, and a series of indoor seating areas for dining and overlooking the park. Defining an urban edge, the auditorium is a prominent, but intimately scaled, multi-use venue for both corporate and public events. Although nestled into the landscaped park, the building's strong

"This beautiful tower has rightly become a new symbol of Oklahoma City, bringing life and regeneration back to its central business district."

David Scott, Juror, Laing O'Rourke

Jury Statement

This building is remarkable, not only for its inherent design excellence, but for the seeming unlikelihood of its location. The culmination of a concerted effort to reduce sprawl and consolidate activity in a newly resurgent city center, Devon Energy Tower interacts with the nearby Myriad Gardens, providing new levels of public engagement in the area. The three-sided shape of the tower, visible from all angles, reinforces its centrality in the city's life and public image.

presence activates street life and supports the downtown's vitality, while providing dramatic views of downtown and Myriad Botanical Gardens.

The building is among the ten largest LEED-NC Gold-certified buildings worldwide. The chosen site has direct access to public transit. Construction of the building minimally impacted the natural environment by diverting 68,000 tons (61,669 metric tons) of waste and concrete from landfills and recycling 100 percent of a demolished parking deck on the site. The building also performs well operationally. Potable water consumption is reduced by 50 percent through landscape design and irrigation; overall water use is reduced by 40 percent. Energy use is modulated by district cooling with on-site cogeneration, personal comfort control for 50 percent of occupants and personal lighting control for 90 percent of occupants.

Finalist
Best Tall Building Americas

Tree House Residence Hall
Boston, United States of America

Completion Date: June 2012
Height: 85 m (280 ft)
Stories: 21
Area: 13,519 sq m (145,517 sq ft)
Use: Residential
Owner: Massachusetts State College Building Authority
Architect: ADD Inc
Structural Engineer: Odeh Engineers, Inc.
MEP Engineer: WSP Flack + Kurtz
Main Contractor: Suffolk Construction
Other Consultants: C3 (code); Ground (landscape); Lerch Bates (vertical transportation); Nitsch Engineering (civil); SGH (façade)

"Art and Architecture have combined collaboratively to produce a building that makes a sustainable statement for Boston."

Robert Okpala, Juror, Buro Happold

This new residential tower results from a highly unusual collaborative process and responds to the unique living/learning requirements of art school students. Inspired by Gustav Klimt's painting *Tree of Life*, this innovative high-rise includes 493 beds for freshmen and sophomores in 136 suites configured in one-, two-, or three-bedroom layouts. The building features a ground-floor café and living room, a second-floor health center, and a "Pajama Floor" at the third level with communal kitchen, game room, laundry facilities, and fitness center. Studio spaces alternate with lounges on the 17 upper floors.

The new tower is located along Boston's Huntington Avenue, in a heterogeneous neighborhood of warm-toned, brick buildings, residence halls, and academic facilities. The MBTA's Green Line passes directly in front of the site and the Colleges of the Fenway path bounds the southern edge. The project's curved stone base accommodates an underground tunnel that

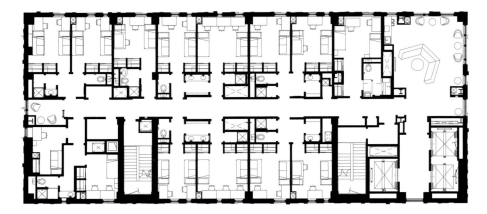

> "The façade composition is a beautiful abstraction of tree bark, an appropriate 'skin' for a building named the Tree House Residence Hall."
>
> Richard Cook, Juror, COOKFOX Architects

swerves through the site and required architects to cantilever the rectangular building above.

During the design process, the team worked to harmonize the goals and aspirations of professors, administrators, students, trustees, alumni, city and state agencies, neighbors, and the building's owner. The architect conducted in-depth benchmarking, hosted focus groups and an 85-person design charrette, and developed full-scale mock-up units for students to experience and critique. Students in the college's architecture and interior design programs helped shape some of the project's common areas, including the ground floor café. Lean construction methods were used to fast-track building trades and bring the project to completion three months before the Fall opening.

The exterior is an organic mosaic of over 5,000 composite aluminum panels of varying depths and hues. Dark browns at the base mirror tree bark before growing progressively lighter to make the building

Previous Spread
Left: Overall view of tower from the west
Right: Close-up view of the façade

Current Spread
Opposite Top: Typical floor plan
Opposite Bottom: Overall view from the street
Right: Interior view of lounge
Below: Main lobby

Jury Statement

Created on a limited public-university budget and an aggressive schedule, the Tree House represents a well-considered solution that announces itself as a foundry of inspiration and innovation. It achieves environmental conservation goals while still providing a memorable environment for the young minds inhabiting the building. It is indeed notable that a project borne of an intensive series of consultations and adjustments can result in such an individualistic, singular statement about its community.

appear taller and lighter in the skyline. Green window panels punctuate the façade like the leaves of a tree.

The project's interior spaces are infused with art ranging from commissioned alumni pieces in the lobby to a rotating gallery on the third floor. While the budget did not allow for expensive finishes, designers drew on the possibilities of modest materials such as carpet and paint to develop a bold visual statement that activates the space through color.

The residence hall's design and engineering decisions were made with solar orientation in mind. Windows on the tower's north side provide light favorable to the work of resident art students, while the smaller number of windows on the south side help reduce heat. The windows are operable and the school employs an electronic system that informs students of advisable times to open or close them.

The building received a Silver LEED certification from the U.S. Green Building Council and its energy usage is 22 percent more efficient than code mandates. Other green features include double-insulated metal panels, and low-flow plumbing fixtures that reduce the amount of potable water usage by 33 percent. More than 50 percent of the material used in the residential hall has recycled content, 20 percent from local sources, and 70 percent of the wood is certified by the Forest Stewardship Council.

Nominee
Best Tall Building Americas

1214 Fifth Avenue
New York, United States of America

This project addresses the urban context of the Upper East Side on multiple levels, with a carefully composed massing of five interlocking forms. The tower animates the skyline with a varied silhouette shaped by three setbacks. The building program and superstructure are integral to one another. The uses of below-grade parking, base-level medical offices, and tower-level apartments correspond to the arrangement of structural materials. The steel framing of the base spans the massive mechanical spaces that also serve the adjacent cancer research center. The cores of the concrete residential tower above also house the 500-foot (152-meter) central chimneys necessary for the medical spaces below.

Designed to attain LEED Silver, the building uses 30 percent less water and is 15 percent more energy efficient than code, while the tower's eight-inch flat-plate concrete slab was designed to use 30 percent less concrete than mandated. Modifying the slab edge allowed the window wall to be expressed without overly prominent horizontal slab covers, and at 25 percent lower construction cost.

Completion Date: October 2012
Height: 156 m (513 ft)
Stories: 43
Area: 39,750 sq m (427,865 sq ft)
Use: Residential/Office
Owner: Mount Sinai Medical Center
Architect: Pelli Clarke Pelli Architects (design); SLCE Architects (architect of record)
Structural Engineer: WSP Cantor Seinuk
MEP Engineer: Jaros Baum & Bolles
Main Contractor: Gotham
Other Consultants: Israel Berger & Associates (façade)

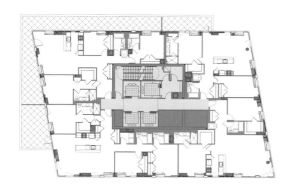

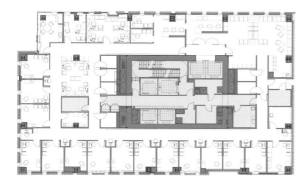

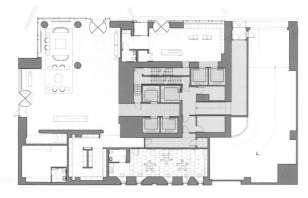

Opposite: Overall view of tower from the north

Above: Floor plans (from top to bottom) – typical residential, typical medical office, and ground floor

Top Right: Aerial view of tower and adjacent cancer research center

Bottom Right: Close-up view of the façade

Nominee
Best Tall Building Americas

Ann & Robert H. Lurie Children's Hospital of Chicago
Chicago, United States of America

The Ann & Robert H. Lurie Children's Hospital of Chicago demonstrates a unique approach to high-rise hospital design. By moving to a neighborhood in downtown Chicago, the client wished to leverage opportunities for collaboration in patient care, teaching, and research with Northwestern University. This location, however, offered some distinct design challenges. At 1.25 million square feet, on a site area of 1.8 acres, the hospital required a creative approach to the building stack.

The results include a number of unique features, including an 11th-floor sky lobby and sky garden, and public amenities on the 11th and 12th floors. Loading docks, in addition to the 27 elevators and an entrance lobby, took up most of the available first-floor area, the remainder being too small to accommodate the space programmed for the Emergency Department, which was assigned to the second floor. This solution improved operational efficiency in a high-rise hospital, through the standardization of processes and work flows.

Completion Date: June 2012
Height: 136 m (447 ft)
Stories: 23
Area: 116,400 sq m (1,253,381 sq ft)
Use: Hospital
Owner: Ann & Robert H. Lurie Children's Hospital of Chicago
Architect: ZGF Architects LLP; Solomon Cordwell Buenz; Anderson Mikos Architects, Ltd.
Structural Engineer: Magnusson Klemencic Associates
MEP Engineer: Affiliated Engineers, Inc.
Project Manager: ARCADIS
Main Contractor: Mortenson/Power, Joint Venture
Other Consultants: CYLA Design Associates (landscape); Lerch Bates (vertical transportation); Mikyoung Kim Design (sky garden design); Ruby+Associates (erection engineer); RWDI (wind); Sako & Associates, Inc. (security); V3 Companies of Illinois (civil); Lassen Associates (technology)

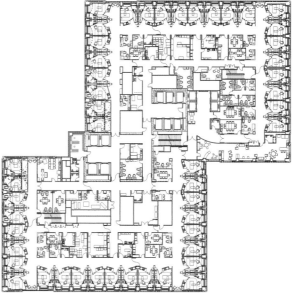

Opposite: Overall view from north
Top Left: 11th floor sky garden
Bottom Left: Context view
Top Right: View looking up at the building's cantilever
Bottom Right: Floor plan – typical acute care unit at level 20

Nominee
Best Tall Building Americas

LOVFT
Santa Catarina, Mexico

Lovft is an apartment building situated just in front of an authentic natural marvel: a millennia-old canyon, known as the Cañón de la Huasteca, located outside Monterrey, Mexico, on the city's west side.

Lovft's main highlight is "The Heart," located on the 22nd level of the building. At this level, there is a lounge, a business center, and a fitness and entertainment center; the overall volume is expressed on the exterior with a dramatic red box. An additional lounge terrace is located on the top floor.

A ventilated façade, roof insulation, low-E double-glazed windows, and the use of efficient lighting and the thermal mass of concrete allow for an annual reduction of 38 percent in total energy use compared to a conventionally constructed building with the same area, orientation, and glazed area. West-facing windows are set back and protected from direct solar radiation by balconies, while projected concrete walls on the south side provide solar control while allowing views and privacy.

Completion Date: February 2012
Height: 132 m (432 ft)
Stories: 32
Area: 26,881 sq m (289,343 sq ft)
Use: Residential
Owner/Developer: ZdC
Architect: Vidal Arquitectos
Structural Engineer: Salvador Aguilar
MEP Engineer: CYMESA
Project Manager: DPS
Main Contractor: Stiva

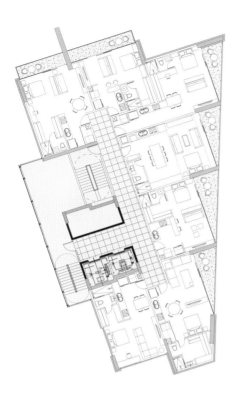
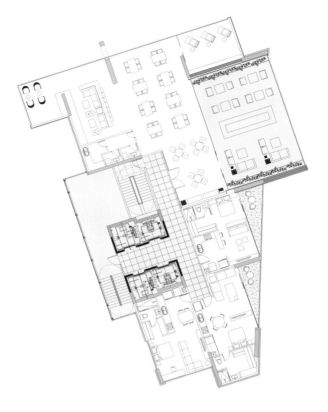

Opposite: Overall view of tower from the southeast
Top: Floor plans – typical residential (left), amenity space and terrace at Level 22 (right)
Bottom Left: View of tower in context
Bottom Right: View from the Level 21 sky terrace

Nominee
Best Tall Building Americas

Mercedes House
New York, United States of America

Located at the western edge of Midtown Manhattan, the Mercedes House development occupies more than half of a city block, and incorporates a variety of commercial and residential programs, including a 55,000-square-foot auto showroom and a horse stable for the NYPD Mounted Police.

The residential form includes 865 units (695 rental units and 170 condo units), including 20 percent affordable housing. The overall massing of the project slopes up and away from De Witt Clinton Park, starting at 86 feet along 11th Avenue and climbing up to 348 feet. Securing light and air for a great majority of apartments, the double-loaded corridor shifts diagonally across the site in a unique orientation to the Manhattan grid, reducing the building's mass adjacent to the neighboring buildings. Each floor steps up from the one below, allowing for unobstructed views to the park and the Hudson River and providing private roof terraces with green roofs on every floor.

Completion Date: November 2012
Height: 106 m (348 ft)
Stories: 32
Area: 120,000 sq m (1,291,669 sq ft)
Use: Residential
Developer: Two Trees Management
Architect: TEN Arquitectos
Structural Engineer: Rosenwasser/Grossman
MEP Engineer: Ettinger Engineering Associates
Main Contractor: Green Star Builders, LLC
Other Consultants: Israel Berger & Associates (façade); Langan Engineering and Environmental Services (geotechnical); Philip Habib & Associates (civil); Viridian Energy & Environmental, LLC (LEED); William Vitacco Associates, Ltd. (expeditor)

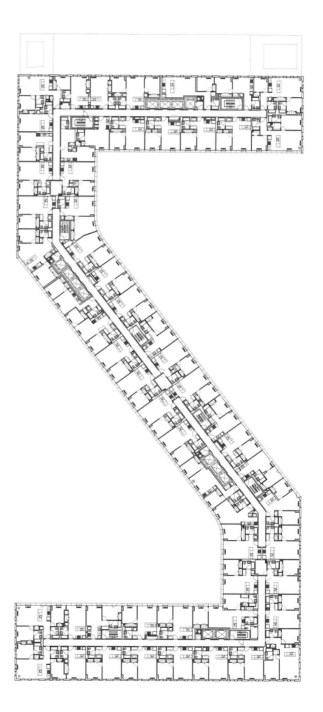

Opposite: Overall view of tower from the west

Above: Views of the amenity space on the podium rooftop

Right: Floor plan – typical residential at level 5

Nominee
Best Tall Building Americas

Trump International Hotel & Tower
Toronto, Canada

This soaring 60-story tower is the tallest all-residential and the second-tallest building in Canada. The glass tower emerges from a stone-clad "reflection" of the adjacent historic limestone buildings to the west and to the south and is accented by a corner light-sculpture element that extends from the sidewalk to the top of the articulated spire.

The tower's efficiency extends through MEP design to structural design and space planning. It makes use of the Enwave District Heating and Cooling System, which draws cold water from Lake Ontario for use in cooling buildings. Hotel and condominium floors are column-free due to post-tensioned concrete floor spans. Separate car and service access, a porte-cochère, a hotel and condominium lobby, and a retail space are all contained on a 45-by-34-meter site. Car parking is located above grade, between the condominium and below the hotel lobby, and uses "stackers": mechanical devices that allow one car to be elevated above another to make maximum use of the volume.

Completion Date: April 2012
Height: 277 m (908 ft)
Stories: 63
Area: 74,510 sq m (802,000 sq ft)
Use: Residential/Hotel
Owner/Developer: Talon International Development, Inc.
Architect: Zeidler Partnership Architects
Structural Engineer: Yolles
MEP Engineer: Hidi Rae Consulting
Project Manager: Lewis Builds Corporation; Brookfield Multiplex
Main Contractor: BLT Construction
Other Consultants: HH Angus (vertical transportation); IBI Group, Inc. (landscape); IIBYIV (interiors); J.E. Coulter Associates Ltd. (acoustics); Randal Brown and Assocaites Ltd. (code); Speirs and Major (lighting)

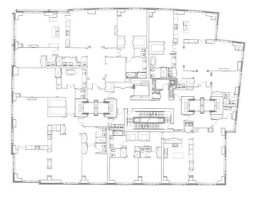
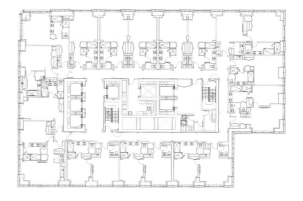
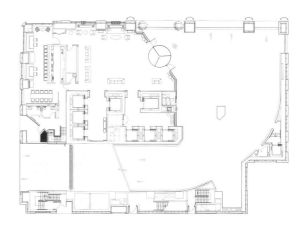

Opposite: Overall view of tower in context from the northwest

Above: Aerial view of tower

Right: Floor plans (from top to bottom) – typical residential, typical hotel, and ground floor

Nominees Best Tall Building Americas

Coast at Lakeshore East
Chicago, United States of America

Completion Date: April 2013
Height: 142 m (467 ft)
Stories: 47
Area: 62,364 sq m (671,281 sq ft)
Use: Residential
Developer: Magellan Development Group LLC
Design Architect: bKL Architecture LLC
Architect: Loewenberg Architects LLC
Structural Engineer: CSAssociates, Inc.
MEP Engineer: Advance Mechanical Systems, Inc.; Gurtz Electric Co.; O'Sullivan Plumbing Inc.
Main Contractor: James McHugh Construction Co.
Other Consultants: dbHMS (LEED); Mackie Consultants, LLC (civil); Wolff Landscape Architecture, Inc. (landscape)

Helicon
San Pedro Garza Garcia, Mexico

Completion Date: January 2012
Height: 156 m (512 ft)
Stories: 33
Area: 40,286 sq m (433,635 sq ft)
Use: Office
Owner/Developer: Vertical Developments
Architect: Vidal Arquitectos
Structural Engineer: SOCSA
MEP Engineer: CYMESA
Main Contractor: DOCSA

Pacifica Honolulu
Honolulu, United States of America

Completion Date: April 2012
Height: 127 m (418 ft)
Stories: 46
Area: 48,001 sq m (516,678 sq ft)
Use: Residential
Owner/Developer: OliverMcMillan
Architect: Architects Hawaii, Ltd.
Structural Engineer: Baldridge & Associates Structural Engineering, Inc.
MEP Engineer: IDS Popov
Main Contractor: Hawaiian Dredging Construction, Co.; Ledcor Construction
Other Consultants: Philpotts Interiors (interiors); Jules Wilson I.D. (interiors)

Reforma 342
Mexico City, Mexico

Completion Date: July 2012
Height: 152 m (499 ft)
Stories: 35
Area: 80,800 sq m (869,724 sq ft)
Use: Office
Owner: Salomon Kamaji; Moises Farca
Developer: Pulso Inmobiliario; MF Ingenieros
Architect: Colonnier y Asociados, S. C.
Structural Engineer: CADAE
MEP Engineer: DYPRO; Garza Maldonado; PIESA
Main Contractor: MF Ingenieros
Other Consultants: DLC (landscape); Front, Inc. (façade); Luz en Arquitectura (lighting)

Nominees Best Tall Building Americas

Rush University Medical Center Hospital Tower
Chicago, United States of America

Completion Date: January 2012
Height: 77 m (252 ft)
Stories: 14
Area: 78,659 sq m (846,678 sq ft)
Use: Hospital
Owner: Rush University Medical Center
Architect: Perkins + Will
Structural Engineer: Thornton Tomasetti
MEP Engineer: Environmental Systems Design, Inc.; IBC Engineering
Main Contractor: Power Jacobs Joint Venture
Other Consultants: Hitchcock Design Group (landscape); Hoerr Schaudt (landscape); Terra Engineering (civil)

Torre Begonias
Lima, Peru

Completion Date: June 2013
Height: 120 m (395 ft)
Stories: 26
Area: 30,112 sq m (324,123 sq ft)
Use: Office
Owner: Urbanizadora Jardin S.A.
Developer: Cubica
Architect: Arquitectonica
Structural Engineer: Antonio Blanco
MEP Engineer: Pronasa – Proterm
Project Manager: Proyecta Ingenieros
Main Contractor: AESA Constructora
Other Consultants: Alejandro Molina (vertical transportation); IBA (façade); Miranda & Nasi (LEED)

Torre Paseo Colón 1
San José, Costa Rica

Completion Date: February 2013
Height: 98 m (322 ft)
Stories: 29
Area: 35,000 sq m (376,736 sq ft)
Use: Residential/Office
Owner: Torres Paseo Colon S.A.
Developer: Grupo Inmobiliario del Parque S.A.
Architect: Arquitectura y Diseño S.A. (design); Ismael Leyva Architects (architect of record)
Structural Engineer: IECA International
MEP Engineer: Circuito S.A.
Main Contractor: Proycon

Best Tall Building
Asia & Australasia

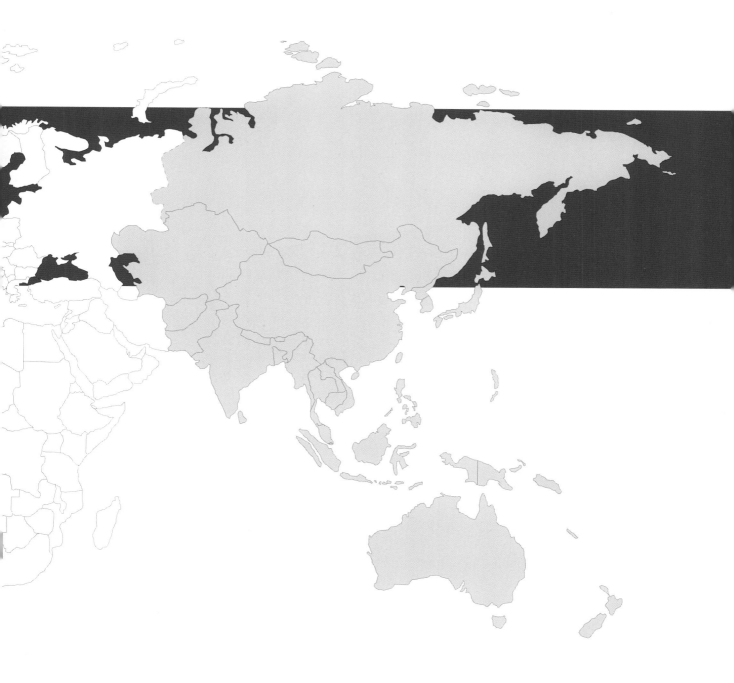

Winner
Best Tall Building Asia & Australasia

CCTV Headquarters
Beijing, China

Completion Date: May 2012
Height: 234 m (768 ft)
Stories: 54
Area: 316,000 sq m (3,401,396 sq ft)
Use: Office
Owner: China Central Television
Developer: General Office of CCTV New Site Construction & Development Program
Architect: OMA (design); ECADI (architect of record)
Structural Engineer: Arup
MEP Engineer: Arup
Main Contractor: China State Construction Engineering Corporation
Other Consultants: DHV Building and Industry (acoustics); Front, Inc. (façade); Inside/Outside (landscape); Lerch Bates (vertical transportation); Lighting Planners Associates Pte Ltd. (lighting)

"The building's complicated building structure, outstanding construction, and unique building function have made it an instant landmark, not only for China, but for the world."

Nengjun Luo, Juror, CITIC Heye Investment

The CCTV Headquarters is an unusual take on the skyscraper typology. Instead of competing in the race for ultimate height and style through a traditional two-dimensional tower soaring skyward, CCTV's loop poses a truly three-dimensional experience, culminating in a 75-meter cantilever.

The building's form facilitates the combination of the entire process of TV making in a loop of interconnected activities. Two towers rise from a common production studio platform, the Plinth. Each tower has a different character: Tower 1 serves as editing area and offices, and Tower 2 is dedicated to news broadcasting. They are joined by a cantilevering bridge for administration, the Overhang.

The main lobby, in Tower 1, is an atrium stretching three floors underground, and three floors up. It has a direct connection with Beijing's subway network, and is the arrival and departure hub for the 10,000 workers inside CCTV Headquarters. Connected to the lobby, 13

 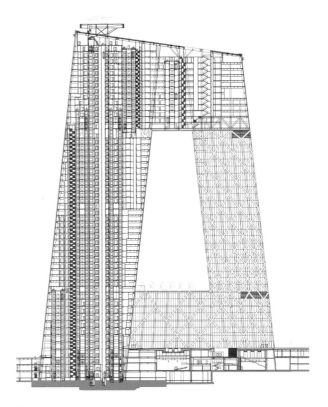

production studios (the largest is 2,000 square meters) perform the main function of the building: TV making.

The building also facilitates an unprecedented degree of public access to the production of China's media: a Public Loop takes visitors on a dedicated path through the building, revealing everyday studio work as well as the history of CCTV, and culminating at the edge of the cantilever, with spectacular views towards the CBD, the Forbidden City, and the rest of Beijing. A Media Park

forms a landscape of public entertainment, outdoor filming areas, and production studios as an extension of the central green axis of the CBD.

The innovative structure of CCTV is the result of long-term collaboration between European and Chinese architects and engineers to achieve new possibilities for the high-rise. Early on, the team determined that the only way to deliver the desired architectural form was to engage the entire façade structure, creating in essence an

Previous Spread
Left: Aerial view
Right: View of tower in context

Current Spread
Opposite Top: Time lapse of tower construction
Opposite Bottom Left: Interior view
Opposite Bottom Right: Typical section
Right: Looking down from a portal window in the observation deck

external continuous tube system. The tube, which resists all of the lateral forces on the building and also carries much of the gravity force, is ideally suited to deal with the nature and intensity of permanent and temporary loading on the building.

The engineering forces at work are thus rendered visible on the façade: a web of triangulated steel tubes – diagrids – which, instead of forming a regular pattern of diamonds, become dense in areas of greater stress and looser and more open in areas requiring less support. The façade itself becomes a visual manifestation of the building's structure.

The structural system is a versatile, efficient structure that bridges in bending and torsion between the Towers to create the continuous form of the Overhang section, providing enough strength and stiffness in the Towers to carry loads to the ground. The structural system stiffens the podium and tower bases to favorably distribute loads to the foundation. It enables performance to be

"The CCTV building is the type of building that may not happen again. It is an incredible achievement in terms of structural engineering and iconography; in some ways it is the Eiffel Tower of our time."

Jeanne Gang, Jury Chair, Studio Gang Architects

optimized, through adjustment of the bracing pattern, to satisfy contrasting demands of stiffness and flexibility.

The structural system also provides maximum flexibility for the bespoke planning of the interiors, since bracing is not needed within the floor plates. This allows large studio spaces to be laid out within the towers. It has enabled the Overhang section to be constructed without the need for temporary propping, since the braced skin provided stability as the steelwork was cantilevered out from the towers. This type of structure has a high degree of inherent robustness and redundancy, due to the potential for adopting alternative load paths in the unlikely event a key element is removed.

The self-supporting hybrid façade structure features high-performance glass panels with a sun shading of 70 percent open ceramic frit, creating the soft silver-grey color that gives the building a surprisingly subtle presence in the Beijing skyline.

Opposite: View looking up at the cantilever
Right: Street view in context
Below: Floor plans – level 41 (top) and level 15 (bottom)

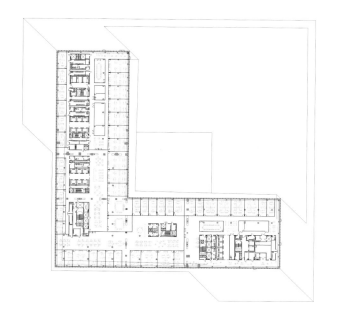

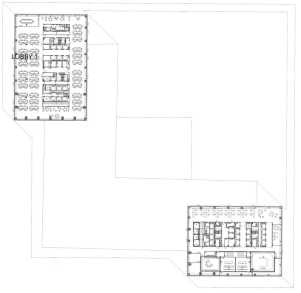

Jury Statement

Conflating expectations of what a skyscraper is, and can or should do, the CCTV Headquarters has now become embedded in the thought process of the making of tall buildings. It singlehandedly paved the way from the height-obsessed, set-back skyscraper of the past to the sculptural and spatial skyscraper of the present, at the scale of the urban skyline.

Its stunning form, which appears both powerful and conflicted, as if pulled in several directions, symbolizes the multiple functions of the program and the dynamic positioning of its nation on the world stage. The unique architectural design contrasts significantly with historical building styles in Beijing, yet it could never be classified as a homogenizing force.

As a piece of structural engineering, CCTV is also an object lesson for those who wish to push the boundaries and sweep aside the received notions of skyscraper design. The building's design violates conventions while validating and rewarding intensive and focused collaboration and study.

Finalist
Best Tall Building Asia & Australasia

C&D International Tower
Xiamen, China

Completion Date: May 2013
Height: 219 m (720 ft)
Stories: 49
Area: 83,066 sq m (894,115 sq ft)
Use: Office
Owner/Developer: Xiamen C&D Corporation Limited
Architect: Gravity Partnership Ltd. (design); Shanghai Institute of Architectural Design & Research (architect of record)
Structural Engineer: Shanghai Institute of Architectural Design & Research
MEP Engineer: Shanghai Institute of Architectural Design & Research
Project Manager: Xiamen C&D Real Estate Co., Ltd
Main Contractor: China Construction Third Engineering Bureau Corp., Ltd.
Other Consultants: Arup (façade); Gravity Green Ltd. (landscape); HASSELL (interiors)

"External shading is used to great effect; connecting the tower and the low-rise commercial development that surrounds it and creating a dramatic integrated building complex."

David Scott, Juror, Laing O'Rourke

The C&D complex sits on the most prominent location of the future Central Business District of Xiamen. Situated immediately in front of the coastline, the complex enjoys a direct connection with the seashore. The building masses are arranged in an open block formation, revealing views to the sea for the sites beyond, and forming a pedestrian network between the building masses.

Rather than creating a large podium retail block, the commercial podium was broken down into smaller blocks forming naturally shaded internal streets, which is in character with the street alleys that are common to the urban architecture of Southern China. The sunken retail arcade opens up to a central plaza surrounded by the buildings, bringing in natural light and ventilation to the vast underground pedestrian network which connects to nearby commercial activities, while helping pedestrians reach the promenade along the coastline. This multilevel pedestrian network also acts as the main

Previous Spread

Left: Overall view from the southwest

Right: Close-up view of the façade with louvers

Current Spread

Left: West elevation showing building's tapered form and operable windows

Opposite Left: Typical office floor plan

Opposite Right: Overall view from the northeast

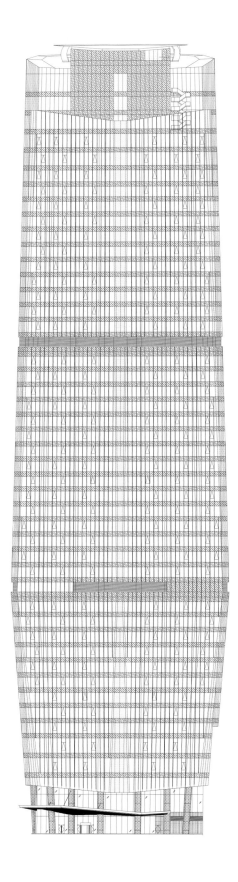

Jury Statement

The C&D International Tower anchors a comprehensive project, with great potential to form the new Central Business District for Xiamen. With relatively slight tapering, this building achieves a compelling sculptural presence on the skyline while affording ample accommodation at the ground level. The designers executed a clear strategy around limiting urban heat islands, supporting greenery, and channeling natural breezes and daylight to the advantage of the tower and the CBD as a whole.

axis of the site, dividing the 200-meter-long site into more easily accessible zones.

The complex consists of a 49-story main office tower, a seven-story podium shopping mall connected to the tower at its upper floors, and two freestanding store buildings. The office tower will be the headquarters of the Xiamen C&D Corporation, the largest enterprise in Fujian province.

The iconic building form represents the enterprise with efficient internal spaces as the first priority in the design consideration. The architectural design of the main tower breaks from the tradition of regular rectangular form. By tapering at both ends the tower's sculptural form is exaggerated while creating more open space at the ground floor in the tight rectilinear site. The multifaceted elevations also break down the mass of the skyscraper volume, harmonizing its relationship with the attached shopping mall form. The striking sculptural geometry of the tower is realized without compromising

"This tower skillfully utilizes passive green strategies, including naturally shaded internal streets which capitalize on the sea breezes to cool the retail space."

Richard Cook, Juror, COOKFOX Architects

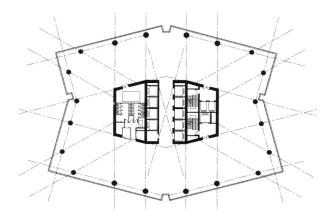

regular floor plans, as the core walls are angled in plan to be parallel to the angles of the façade.

A passive solar design principle was adopted as the main strategy throughout the building design process. The project site is rectangular in shape, with its long edges running north to south, thus exposing most of the building façade to the east and west sunlight.

The east- and west-facing façades of the tower and podium are equipped with vertical fins to help shade the building from direct sunlight. All insulated glass unit (IGU) panels are low-E coated to minimize energy loss for the interior space while allowing maximum natural daylight penetration. Operable windows provide natural ventilation and limit dependence on mechanical air conditioning. Clear floor-to-ceiling height was maximized to allow optimal daylight penetration deep into the floor plate. These moves helped the building achieve high energy efficiency.

Finalist
Best Tall Building Asia & Australasia

PARKROYAL on Pickering
Singapore

"Looking up (or down) at PARKROYAL's terraced greenery induces yet another spell-binding moment of wonder from the high-rise ouevre of WOHA."

Antony Wood, Juror, CTBUH

Completion Date: January 2013
Height: 89 m (292 ft)
Stories: 16
Area: 20,648 sq m (222,253 sq ft)
Use: Hotel
Owner/Developer: UOL Group Limited
Architect: WOHA
Structural Engineer: TEP Consultants Pte Ltd.
MEP Engineer: BECA Carter Hollings & Ferner Pte Ltd.
Main Contractor: Tiong Seng Contractors Pte Ltd.
Other Consultants: CCW Associates Pte Ltd (acoustics); Lighting Planners Associates (S) Pte Ltd. (lighting); LJ Energy Pte Ltd. (sustainability); Meinhardt (façade); Rider Levett Bucknall LLP (quantity surveyor); Tierra Design (S) Pte Ltd. (landscape)

PARKROYAL on Pickering is a hotel in the midst of Singapore's high-density city center. A contoured podium responds to the street scale, drawing inspiration from terraformed landscapes, such as rice paddies. These contours create dramatic outdoor plazas and gardens, which flow seamlessly into the interiors. Greenery from nearby Hong Lim Park is drawn up into the building in the form of planted valleys, gullies, and waterfalls.

The podium houses the above-ground car park, transforming it into a sculptural urban object: its roof becomes a lush landscaped terrace, housing the hotel's recreational facilities, which include birdcage-shaped cabanas.

Above the podium the crisp and streamlined tower blocks harmonize with surrounding high-rise office buildings. They are attenuated into an open-sided courtyard configuration, breaking down the "wall-of-buildings" effect and maximizing views and daylight.

Finalist
Best Tall Building Asia & Australasia

Pearl River Tower
Guangzhou, China

Completion Date: April 2013
Height: 309 m (1,015 ft)
Stories: 71
Area: 165,840 sq m (1,785,087 sq ft)
Use: Office
Owner/Developer: The Guangzhou Pearl River Tower Properties Co., Ltd.
Architect: Skidmore, Owings & Merrill LLP (design); Guangzhou Design Institute (architect of record)
Structural Engineer: Skidmore, Owings & Merrill LLP
MEP Engineer: Skidmore, Owings & Merrill LLP
Main Contractor: Shanghai Construction Group
Other Consultants: SWA Group (landscape); Pivotal (lighting); Fortune Consultants, Ltd. (vertical transportation); Highrise Systems, Inc. (maintenance); Rolf Jensen & Associates, Inc. (fire); RWDI (wind); Shen Milsom Wilke, Inc. (acoustics)

"The Pearl River Tower has a steadfast environmental approach and advances the idea of integrated renewable energy in high-rise buildings."

Jeanne Gang, Jury Chair, Studio Gang Architects

Using some of the most sophisticated technologies currently available, the designers of Pearl River Tower created a highly integrated structure that derives its efficiencies by applying previously tested solutions in a combination never before accomplished at such at large scale. It was important to both the client and the design team that a holistic approach be used, so as to avoid an array of solutions that might be conceptually compelling, but would not survive the rigors of design development and future value-engineering exercises. This demanded a design approach that was not form driven, but performance based, with all systems having a degree of interdependency.

Thus, the building has been carefully shaped to use natural forces to maximize its energy efficiency. The tower's sculpted body directs wind to a pair of openings at its mechanical floors, pushing turbines that generate energy for the building. East and west elevations are straight, while the south façade is concave; the north façade is convex. The south side of the building is

Previous Spread
Left: Overall view of tower
Right: Glass laminated photovoltaic panels on the roof of the tower

Current Spread
Left: Night view
Opposite Left: Rendered view of one of the openings that house the wind turbines
Opposite Right: Building section

dramatically sculpted to direct wind through the four openings, two at each mechanical level.

The building's siting and evocative curving shape work together to drive performance. Its generally rectangular floor plate has been shifted slightly from Guangzhou's orthogonal grid in order to maximize use of prevailing breezes, and to better capture the sun's energy through the strategic location of photovoltaics.

The tower's shading system uses automated, daylight-responsive blinds set within the building's double-skin façade, thereby reducing the building management's operational needs. Its ventilation/dehumidification system uses heat collected from the double-skin façade as an energy source.

The integrated façade assembly provides very good thermal performance, as well a high level of natural daylight to the space. Low-energy, high-efficiency lighting systems use radiant panel geometry to assist in the distribution of light. The double-skin façade also allows greater flexibility in the layout of office space, as it reduces the amount of internal mechanical chases required for ventilation, heating, and cooling.

The tower's mechanical design approach also allowed the architects to reduce the building's floor-to-floor height from 4.2 meters to 3.9 meters, reducing the number of constructed stories by five. Occupants can be comfortably positioned close to perimeter walls. The radiant cooling, chilled ceiling and decoupled ventilation system provide improved human thermal comfort, efficient heat exchange, and improved office acoustics.

Jury Statement

Pearl River Tower's highly integrated architectural, structural, and mechanical solutions are impressive. The building's siting and its evocative curving shape are performance driven, and represent an example of a 21st century tower that responds responsibly to local climactic conditions and global energy concerns. Sculpted to both catch the wind and entice the eye, this building manages to achieve iconicity for its city without proclaiming its green credentials in a superfluous manner.

The ventilation system is delivered via a raised access floor, providing improved indoor air quality and air change effectiveness. There is also a reduced cost of tenant fit-out and future retrofits due to the absence of fan coils, VAV boxes, filters, ductwork, insulation, and other items typically requiring tenant-specific alterations.

> *"Pearl River Tower has managed to effectively integrate technology and engineering systems to produce a building that is far greater than the sum of its parts."*
>
> Robert Okpala, Juror, Buro Happold

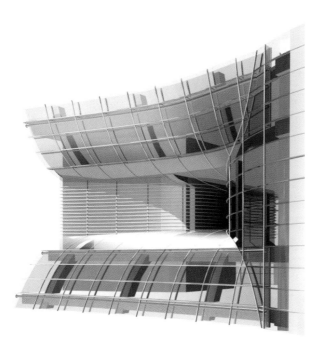

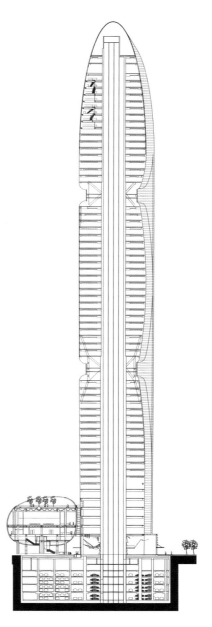

While it is the combination of performance-driven curving shape and exposed vertical-axis wind turbines that fuse Pearl River Tower into the public perception of the Guangzhou skyline, its most significant impact is drawn from the level of integration between sustainable design elements. The combination of turbines, shading systems, a double-skin façade with energy-efficient lighting, ventilation, and mechanical design all work together complementarily, resulting in a substantial decrease in the amount of electrical power required to operate the building's HVAC and lighting systems. Full implementation of Pearl River Tower's sustainable strategies will result in overall energy savings of approximately 30 percent as compared to a conventionally designed building of the same scale, constructed to conform to the Chinese baseline energy code.

Finalist
Best Tall Building Asia & Australasia

Sliced Porosity Block
Chengdu, China

"First with Linked Hybrid in Beijing, and now here with Sliced Porosity, Holl is developing a clear language for the interconnected, multi-tower urban insertion."

Antony Wood, Juror, CTBUH

In the center of Chengdu, China, at the intersection of the first Ring Road and Ren Ming Nam Road, the Sliced Porosity Block forms large public plazas with a hybrid of different functions. The program consists of five towers with offices, serviced apartments, retail, a hotel, cafés and restaurants, and a large urban public plaza. Creating a metropolitan public space, this 278,709-square-meter project takes its shape from the need to distribute natural light, such that each of the residential apartments experiences a minimum of two hours of sunlight each day.

The required minimum sunlight exposures to the surrounding urban fabric prescribe precise geometric angles that slice the exoskeletal concrete frame of the structure. The building structure is white concrete, organized in six-foot-high openings with earthquake diagonals as required, while the "sliced" section's façades are glass. The structural system used in the Sliced Porosity block consists largely of exoskeletal concrete framing, up to 123 meters high at the office towers.

Completion Date: November 2012
Height: Office Block 1 & 2: 123 m (404 ft); Hotel Block: 119 m (389 ft); Apartment Block 1 & 2: 112 m (369 ft)
Stories: Office Block 1 & 2: 29; Hotel Block: 34; Apartment Block 1 & 2: 33
Area: 164,380 sq m (1,769,372 sq ft)
Use: Office Block 1 & 2: Office/Retail; Hotel Block: Hotel; Apartment Block 1 & 2: Residential/Retail
Owner: CapitaLand China
Architect: Steven Holl Architects (design); China Academy of Building Research (architect of record)
Structural Engineer: China Academy of Building Research
MEP Engineer: Arup
Main Contractor: China Construction Third Engineering Bureau Corp., Ltd.
Other Consultants: L'Observatoire International (lighting); Davis Langdon & Seah (quantity surveyor)

The concrete façades express the buildings' structural behavior, as concrete diagonals cut across the column/beam grid where needed to carry vertical and seismic loads. The density and placement of these diagonals follow the requirements of the buildings' carvings and multistory cantilevers. The concrete mix contains a high proportion of recycled materials. A "break" from the concrete and glass material palette occurs in the public History Pavilion, which is faced in raw bamboo and COR-TEN steel.

Allowing the structure to be placed on the façades maximizes floor efficiency and allows large column-free spaces in the buildings. In Chengdu weather, it also serves a sustainable purpose as it acts as a container of thermal mass, conserving the cool indoor temperatures in the summer and warm indoor temperatures in the winter.

The large public space framed in the center of the block is formed into three valleys inspired by a poem of the city's greatest poet, Du Fu (713-770), who wrote, "From

> *"This family of linked buildings creates meaningful space and shapes the way daylight enters the public realm in ways that would be impossible with a traditional point tower."*
>
> Richard Cook, Juror, COOKFOX Architects

Previous Spread

Left: View of the complex from plaza

Right: Active public space at the center of the complex

Current Spread

Opposite Top: Close-up view of the Light Pavilion within the façade

Opposite Bottom: View from within the large central plaza

Right: Overall view of complex from the west

Below: Site plan

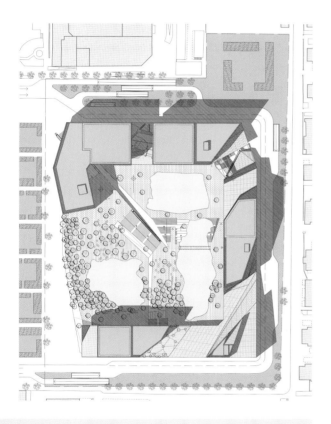

Jury Statement

The buildings of Sliced Porosity Block, with their carefully shaped exteriors, bring the environmental needs of today into the cultural framework of local history. Despite its large scale, the project resists the temptation to create object-icon skyscrapers, instead shaping itself to optimize the distribution of natural light and create a public space with human-scaled moments. The design and spatial *parti* alludes to the region's ancient roots in poetry and nature, yet it also comfortably hosts assertively modern art.

the northeast storm-tossed to the southwest, time has left stranded in Three Valleys." The three plaza levels feature water gardens based on concepts of time. These are the Fountain of the Chinese Calendar Year, Fountain of Twelve Months, and Fountain of Thirty Days. These three ponds function as skylights for the six-story shopping precinct below. Visitors transit between the levels of the public plaza via several means, including an inclined moving sidewalk and shallow stairs.

The designers achieved human scale in this metropolitan rectangle through the concept of "micro urbanism," in which double-fronted shops open to the street as well as the shopping center. Three large openings are sculpted into the mass of the towers as the sites of the Pavilion of History, designed by Steven Holl Architects, the Light Pavilion by Lebbeus Woods, and the Local Art Pavilion.

The Sliced Porosity Block is heated and cooled with 468 geothermal wells, and the large ponds in the plaza harvest recycled rainwater, while the natural grasses and lily pads create a natural cooling effect. High-performance glazing, energy-efficient equipment, and the use of regional materials are among the other methods employed to reach the LEED Gold rating.

Nominee
Best Tall Building Asia & Australasia

Brookfield Place
Perth, Australia

The tower embraced the aspirations of the building's owner to provide a flexible collaborative workplace for its occupants. Knowledge sharing and collaboration underpinned the design philosophy. This approach resulted in a side-core commercial tower with complete visual connectivity across the open-plan floors, which is further reinforced by the vertical connections through generous stair and void zones. This offset core further protects the enclosed space by buffering the façade against the adverse northern sun and the associated solar heat gain.

The site contained five dilapidated historic buildings that had to be restored, upgraded, and integrated into the design. The historic buildings connect with the community and create a human-scaled, urban public realm that has given the site a range of textures and spatial characteristics to enrich the ground plane and create a dynamic mixed-use precinct into which the tower has been carefully inserted.

Completion Date: August 2012
Height: 234 m (769 ft)
Stories: 45
Area: 87,500 sq m (941,842 sq ft)
Use: Office
Owner: Brookfield Office Properties
Developer: Brookfield Multiplex
Architect: HASSELL; Fitzpatrick + Partners
Structural Engineer: Aurecon
MEP Engineer: Aurecon
Main Contractor: Brookfield Multiplex

Opposite: Overall view from the west
Top Left: Active mixed-use ground plane integration
Bottom Left: Preserved historic street façade
Top Right: Typical floor plan with offset core and inter-floor stairs/voids
Bottom Right: Context view in the skyline

Nominee
Best Tall Building Asia & Australasia

Hangzhou Civic Center
Hangzhou, China

The essential core of Hangzhou Civic Center's architecture is a human-scale-oriented garden which is open to the public, surrounded by a tight ring of high-rises which are linked by a two-story skybridge 85 meters above the ground. Inspired by the imagery of the neighboring mountains, the cluster will become part of the unique urban topology and be recognizable from all viewpoints within the city.

The complex uses a mega-block structure composed of six 100-meter high-rises and four podium buildings. It is located on the city axis with the West Lake to the north and Qiantang River to the south. This mega-project accommodates complex functions, including administrative and commercial offices, a conference center, public library, youth activity center, exposition hall, civic service center, and ancillary facilities such as restaurants, supermarkets, and parking. These elements are connected on the basement floor by an eight-meter-wide corridor, which also leads to the Metro and neighboring blocks.

Completion Date: June 2012
Height: 110 m (360 ft)
Stories: 26
Area: 226,200 sq m (2,434,797 sq ft)
Use: Office
Owner/Developer: Hangzhou Qianjiang New City Construction Headquarters
Architect: Tongji Architectural Design (Group) Co., Ltd.
Main Contractor: Zhejiang Construction Engineering Group Co., Ltd.; Zhejiang Greatwall Construction Group Co., Ltd.
Other Consultants: Architectural Design & Research Institute of Zhejiang University (interiors); Dongliang Light (lighting); Shenzhen King Façade (façade); Wuhan Lingyun (façade); Zhongshan Shengxing (façade)

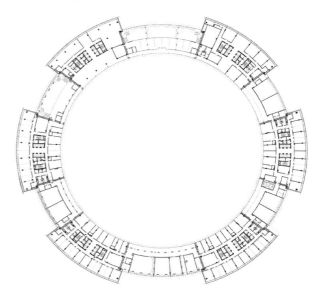

Opposite Top: Overall view of the complex from the south
Opposite Bottom: Context view from nearby opera house and conference center plaza
Above: View looking up at the skybridge ring connections
Top Right: Typical floor plan at skybridge level
Bottom Right: View of the central public garden

Nominee
Best Tall Building Asia & Australasia

Hysan Place
Hong Kong, China

Hysan Place creates outdoor public space on a small site at different levels in a super-dense urban habitat. Its expressive combination of vibrant mixed uses and its porous, yet interconnected, form is a solution to the particular challenge of adding vertical density in a diverse and nuanced context. Between the lower retail podium and the office tower sit nine floors that were designed with internal flexibility in mind. They can be leased as either larger office floors or retail floors depending on market conditions, and are designed to be accessed by either set of office or retail shuttle lifts. This enables flexibility in use over time.

Replacing an opaque and dated predecessor, Hysan Place is the first LEED Platinum-certified building in Hong Kong. Its multilevel public rooftop gardens cover an equivalent of 47 percent of its site area and bring light and air into the middle sections of the building, providing elevated green oases. The large openings in its form induce breezes, which ventilate and improve the surrounding environment.

Completion Date: August 2012
Height: 204 m (670 ft)
Stories: 36
Area: 66,511 sq m (715,918 sq ft)
Use: Office/Retail
Owner/Developer: Hysan Development Company Limited
Architect: Kohn Pedersen Fox Associates (design); Dennis Lau & Ng Chun Man (architect of record)
Structural Engineer: Meinhardt
MEP Engineer: Parsons Brinkerhoff (Asia)
Main Contractor: Gammon Construction Limited
Other Consultants: ALT Cladding & Design (façade); Arup (sustainability); Benoy (interiors – retail); Davis Langdon & Seah (quantity surveyor); Lighting Planners Associates Pte Ltd. (lighting); Shen Milsom Wilke, Inc. (acoustics)

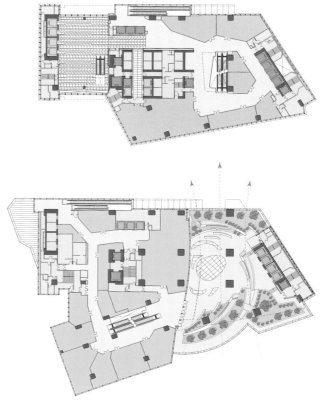

Opposite: Overall view of tower from the south

Top Left: Close-up view of the office sky lobby

Bottom Left: Public rooftop garden on the fourth floor

Top Right: View of tower base

Bottom Right: Floor plans – typical upper level retail at level 9 sky lobby (top), and sky garden at level 4 (bottom)

Nominee
Best Tall Building Asia & Australasia

International Finance Centre
Seoul, South Korea

The International Finance Centre, located at the center of Yeouido Island, serves as a major pedestrian link between the subway station, riverside, and Yeouido Plaza Park. The inspiration for the IFC Seoul came from Asian landscape drawings that often depict very steep, jagged mountains. The four vertical towers were designed to resemble monumental crystalline outcroppings, chiseled to create prismatic forms that accentuate the interplay of light and shadow on the glass façade.

Expansive floor-to-ceiling glass and nearly ten-foot ceilings immerse the office spaces in natural light and provide magnificent, uninterrupted views across Yeouido Park and the Han River to the mountains beyond. Bay depths of more than 42 feet (12.8 meters) from elevator core to the perimeter wall ensure efficient use of space. To improve the experience on subterranean levels, the design includes large skylights to admit natural light, articulating the glass entry pavilions leading into the IFC Seoul Mall.

Completion Date: 2012
Height: One IFC: 186 m (610 ft); Two IFC: 176 m (577 ft); Three IFC: 284 m (932 ft); IFC Hotel: 196 m (643 ft)
Stories: One IFC: 32; Two IFC: 29; Three IFC: 55; IFC Hotel: 38
Area: 505,236 sq m (5,438,315 sq ft)
Use: One IFC, Two IFC, Three IFC: Office; IFC Hotel: Hotel
Owner: AIG Global Real Estate
Developer: AIG Korean Real Estate Development YH
Architect: Arquitectonica (design); Baum Architects (architect of record)
Structural Engineer: Thornton Tomasetti; Chang Minwoo Structural Consultants
MEP Engineer: Cosentini Associates; Woowon M&E
Project Manager: GS Consortium
Main Contractor: GS Engineering & Construction

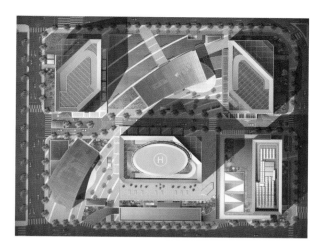

Opposite: Overall view of complex from the northeast
Top Left: Site plan of towers in relation to one another, the tallest, Three IFC is in the center
Bottom Left: The faceted towers reflect Asian landscape drawings of jagged mountains
Top Right: Entrance to the complex
Bottom Right: Aerial view of towers in site context

Nominee
Best Tall Building Asia & Australasia

I Tower
Incheon, South Korea

The tripartite concept of the I-tower started with its three primary occupiers – Incheon Free Economic Zone (IFEZ) headquarters, a UN office, and a rental office zone. The corresponding three basic program volumes are the office tower, the culture wing, and public service wing. Six-story atria are oriented in four different directions, and exterior louvers can be adjusted to match solar conditions. The inverse-sloped atrium provides varied floor plans, different activity spaces and varied conditions for rental office users.

The volume of the building has a pyramid shape sliced out around the corner, and gives a triangular cladding line to the elevations. The upside-down pyramidal mass on the top of the building is a major feature of the project, accommodating the directors of IFEZ and VIP receptions. Underneath the mass is the observation deck and roof garden for all users. In order to achieve the spacious open area with its antigravity slope shape, a mega-truss structure was required.

Completion Date: February 2013
Height: 151 m (494 ft)
Stories: 33
Area: 46,177 sq m (497,045 sq ft)
Use: Office
Owner/Developer: Incheon Free Economic Zone Authority
Architect: Haeahn Architrecure, Inc. (design); Designcamp Moonpark dmp (architect of record); Gyungsung Architects & Engineers (architect of record); TCMC Architects & Engineers (architect of record)
Structural Engineer: Chang Minwoo Structural Consultants
MEP Engineer: Sahm-shin Engineers, Inc.; Ilshin Engineers & Construction
Main Contractor: Daewoo Engineering & Construction
Other Consultants: Ace-all (civil); BEYOND Landscape Design Group (landscape); Yungdo Engineers & Consultants (fire)

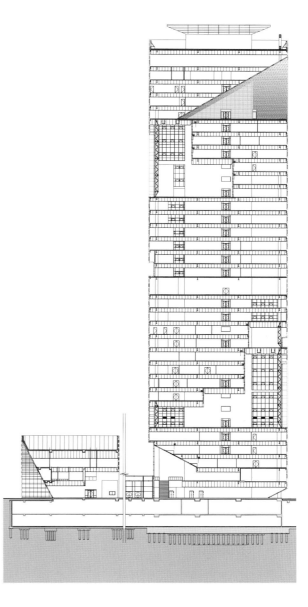

Opposite: Overall view of tower from the northwest
Top Left: Interior view of 29th floor atrium
Bottom Left: View from the roof deck
Top Right: Interior view of concourse
Bottom Right: Typical section

Nominee
Best Tall Building Asia & Australasia

Japan Post Tower
Tokyo, Japan

This new high-rise is built on, and adjacent to, the historic 1931 Central Post Office, next to the busy Marunouchi Station in central Tokyo. It incorporates the existing post office's façade as the base of the building, creating a retail galleria, forming an interface to the urban scale of the adjacent plaza. The tower integrates the lines of the existing building within its façade, creating continuity between the two parts.

It employs a number of sustainable strategies. Rather than ceramic frit coatings on the atrium skylight glass, photovoltaic panels with 10 percent light transmittance were used. This affords control of solar heat gain in the atrium, while absorbed radiation is converted into electricity. The high-performance building envelope integrates a triple-glazed ventilated cavity wall that reduces heat gain in warm seasons and heat loss in cooler seasons. Automated sun shades are located within the cavity to reflect solar radiation and release the heat back into the cavity from where it can be vented out or maintained to reduce heat loss during the winter.

Completion Date: November 2012
Height: 200 m (656 ft)
Stories: 40
Area: 220,000 sq m (2,368,060 sq ft)
Use: Office
Owner: Japan Post Network Co., Ltd.
Architect: JAHN (design); Mitsubishi Jisho Sekkei, Inc. (architect of record)
Structural Engineer: Mitsubishi Jisho Sekkei, Inc.
MEP Engineer: Mitsubishi Jisho Sekkei, Inc.
Main Contractor: Taisei Corporation
Other Consultants: Arup (façade); Werner Sobek Group (façade)

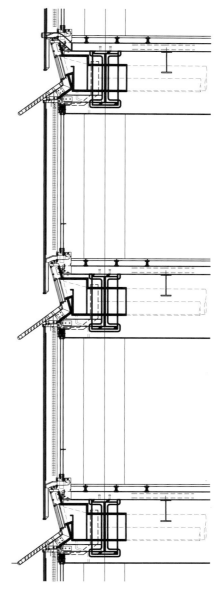

Opposite: Overall view of tower from the northeast
Top: View of base of tower with integrated historic post office building
Bottom Left: Close-up view of the façade corner
Right: Detailed section of west façade double skin

Nominee
Best Tall Building Asia & Australasia

NBF Osaki Building
Tokyo, Japan

The NBF Osaki Building (formerly named Sony City Osaki) houses Sony's R&D department, taking the form of a thin vertical plate. Owing to the narrowness of the building, the offices have flexible, open plans without columns. All the building's mechanisms are integrated into the façades, which were designed in response to the environment. The eastern façade is covered with specialized ceramic louvers that guide rainwater through the system to act as an enormous radiator for cooling the environment operating as an urban "cool spot." This is the first building to install Bioskin, a simple system that circulates rainwater by using electricity generated by solar power through unglazed terracotta pipes arranged along the façade. As the water vaporizes it cools the building and surrounding area.

The building utilizes a seismic-isolation structure which is not common in high-rise design because of the effects of swaying in strong winds. However, lock mechanisms were used to protect the building in high wind conditions.

Completion Date: March 2011
Height: 133 m (436 ft)
Stories: 25
Area: 123,041 sq m (1,335,166 sq ft)
Use: Office
Owner: Nippon Building Fund, Inc.
Architect: Nikken Sekkei, Ltd.
Structural Engineer: Nikken Sekkei Ltd.
MEP Engineer: Nikken Sekkei Ltd.
Main Contractor: Kajima Corporation

Opposite: Overall view of tower the from the west
Top Left: Tenants enjoy the balconies behind the terracotta pipes
Bottom Left: View of the seismic-isolation structural system
Top Right: Overall view of tower from the northeast
Bottom Right: The Bioskin façade as seen from a balcony

Nominee
Best Tall Building Asia & Australasia

Shenzhen Stock Exchange
Shenzhen, China

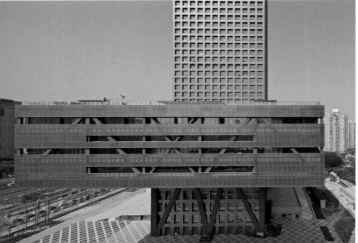

The Shenzhen Stock Exchange is a Financial Center with civic meaning. Located in a new public square, it engages the city not as an isolated object, but as a building to be reacted to at multiple scales and levels. The building's façade wraps the robust exoskeletal grid structure supporting the building in patterned glass. It forms a "deep façade," with recessed openings that passively reduce the amount of solar heat gain entering the building.

SZSE's raised podium is a three-story cantilevered platform floating 36 meters above the ground, with an area of 15,000 square meters per floor and an accessible landscaped roof. This raised podium contains all the Stock Exchange functions, including the listing hall and all stock exchange departments. Lifting the Stock Exchange off the ground has vastly increased its exposure through its elevated position and frames views of the city. But it also liberates the ground level and creates a generous public space beneath what is typically a secure, private building type.

Completion Date: May 2013
Height: 246 m (806 ft)
Stories: 46
Area: 170,000 sq m (1,829,865 sq ft)
Use: Office
Other Use: Stock Exchange
Owner/Developer: Shenzhen Stock Exchange
Architect: OMA (design); Shenzhen General Institute of Architectural Design and Research Co., Ltd. (architect of record)
Structural Engineer: Arup
MEP Engineer: Arup
Main Contractor: General Construction Company of CCTEB
Other Consultants: DHV Building and Industry (acoustics); Inside/Outside (landscape); L&B Quantity Surveyors (quantity surveyor)

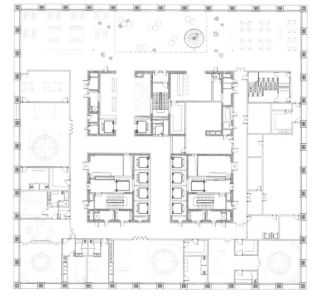

Opposite Top: Overall view of tower from the northwest

Opposite Bottom: The raised podium leaves generous public space below

Top: View of rooftop garden

Bottom Left: Close-up view of the façade showing deep recessed openings

Bottom Right: Floor plan – typical office at level 45

Nominee
Best Tall Building Asia & Australasia

Shibuya Hikarie
Tokyo, Japan

Shibuya Hikarie seeks to become a new high-rise landmark by "stacking up" the diversity and bustle of its surroundings. The building is a multiple use facility accommodating offices, a 2,000-seat theater for musicals, a large 1,000-square-meter event hall and a smaller 300-square-meter event hall, creative spaces, restaurants, shops, and other commercial facilities.

The building's stack consists of blocks whose exteriors are defined by function. The blocks are connected by a common-use deck and sky lobby space, so that office workers will mingle with those using other parts. The resulting synergies will transmit to the exterior the general liveliness of the building as a whole. The subdivision of the building's volume helps to alleviate its rather overwhelming presence against the much smaller structures on its north and south sides. This is reinforced by promoting continuity and circulation between the building and the multiple public transit lines, and with deck areas linking to main streets.

Completion Date: April 2012
Height: 183 m (599 ft)
Stories: 34
Area: 144,546 sq m (1,555,880 sq ft)
Use: Office/Exhibition/Retail
Owner: Council for Promotion of the Shibuya New Cultural District Development Project
Architect: Nikken Sekkei, Ltd. (design); Tokyu Architects & Engineers, Inc. (architect of record)
Structural Engineer: Nikken Sekkei, Ltd.
MEP Engineer: Nikken Sekkei, Ltd.
Main Contractor: Tokyu Construction

Opposite: Overall view of tower from street level
Left: Façade detail highlighting the "stacked box" effect
Top Right: View from theater foyer
Bottom Right: View of tower from pedestrian bridge

Nominee
Best Tall Building Asia & Australasia

Soul
Gold Coast, Australia

Soul is a resort residential tower with ground-level retail, restaurants, and conference facilities located on the beach front. The Soul tower is designed as a passively heated and naturally cross-ventilated building. The tower is single-loaded to maximize views and natural cross-flow ventilation. Its expression is a direct product of its solar orientation. Facing southeast, it is almost entirely transparent. Facing northeast, it is shaded by large concrete columns and balustrades. The north is shaded by concrete blade elements and to the west, openings are minimized and shaded by either concrete balustrades or glass balustrades, reducing the direct solar and thermal load.

The site was previously occupied by an inward-facing retail mall, which failed to engage the beach. The city's only physical connection between its urban fabric and its beach front was a staircase 20 meters wide. Through the Soul project, 130 meters of beach front has been transformed into pedestrianized public domain, activated by restaurants and retail.

Completion Date: July 2012
Height: 243 m (796 ft)
Stories: 76
Area: 49,465 sq m (532,437 sq ft)
Use: Residential/Hotel
Owner/Developer: Juniper Development Group Pty., Ltd.
Architect: DBI Design
Structural Engineer: Glynn Tucker Engineers
MEP Engineer: EMF Griffiths
Main Contractor: Grocon
Other Consultants: C&B Group (urban planner); Windtech (wind); EMF Griffiths (vertical transportation)

Opposite: View of tower from retail base

Top: Overall view of tower from the southeast

Bottom Left: Retail area at base connecting public spaces to the beach

Bottom Right: Typical floor plan

Nominee
Best Tall Building Asia & Australasia

The Ellipse 360
New Taipei City, Taiwan

This slender tower, elliptical in plan, is the symbolic center of a master-planned residential community currently under development on a hillside site overlooking the Tan Shui River, 20 kilometers northwest of central Taipei. Oriented on a view axis that extends across the river toward the revered mountain of Kwan Yin, the tower's clear and simple form serves as a focal point for the surrounding area. The Ellipse's short southwest-northeast axis, through the building's broad face, is oriented toward Kwan Yin, while the long northwest-southeast axis overlooks downtown Taipei.

The expression of the façade is functional and direct: deeply recessed windows bound by continuous terraces edged with sunshades. While the elliptical form is contemporary, the multistory façade of sheltering roof forms is evocative of traditional Taiwanese architecture. Each soffit panel, canted upward from the window line to capture the ambient light of the surrounding site, is warped in plane to mediate between the geometries of window line and balcony edge.

Completion Date: January 2013
Height: 135 m (444 ft)
Stories: 38
Area: 30,421 sq m (327,450 sq ft)
Use: Residential
Owner/Developer: Southern Land Development Co., Ltd.
Architect: Pei Cobb Freed & Partners (design); J.M. Lin Architect, P.C. (architect of record)
Structural Engineer: Columbus Engineering Co., Ltd.
MEP Engineer: Uni Engineering Co., Ltd.
Main Contractor: Chung-Lin General Contractors, Ltd.
Other Consultants: Energydesign Asia (energy concept); George Sexton Lighting Design (lighting); PWP Landscape Architecture (landscape)

Opposite: Overall view of tower from the northeast
Top: View of tower in its hillside context
Left: View from within a balcony
Right: Close-up view of the deep balconies and sunshades

Nominee
Best Tall Building Asia & Australasia

Zhengzhou Greenland Plaza
Zhengzhou, China

The Greenland Plaza tower is instantly recognizable for its sophisticated three- to five-story-tall light-gauge painted aluminum screens. Configured at an outward cant that enhances interior daylighting through scientifically calculated reflections, the screens protect the all-glass exterior from solar gain. The rhythmic cant of the screens, combined with their decreasing size as they rise on the building, creates a dynamic movement that gives the building a fine-grained texture.

The entire experience of the tower is suffused with gradations of light. At the tower's entrance, the circular ground-floor office lobby features floor-to-ceiling windows, offering views of the gardens and the lake beyond. At the 36th floor, the hotel lobby anchors an 18-story atrium; natural light permeates throughout. Above the observation deck, the heliostat – a digitally shaped solar reflector – focuses sunlight into the atrium. Through an intensive series of daylighting studies, the device was specifically designed to maximize the amount of natural light entering the building.

Completion Date: October 2012
Height: 280 m (919 ft)
Stories: 56
Area: 240,169 sq m (2,585,158 sq ft)
Use: Office/Hotel
Owner/Developer: Greenland Group
Architect: Skidmore, Owings & Merrill LLP
Structural Engineer: Skidmore, Owings & Merrill LLP
MEP Engineer: Skidmore, Owings & Merrill LLP
Project Manager: ECADI
Main Contractor: Zhejiang Zhong Tian Construction Group Co., Ltd.
Other Consultants: Beijing Jangho Curtain Wall Co., Ltd. (façade); Fortune Consultants, Ltd. (vertical transportation); Kaplan, Gehring, McCarroll Architectural Lighting (lighting); RJA Group, Inc. (fire); RWDI (wind); Shanghai Liaoshen Curtain Wall Engineering Co., Ltd. (façade); Shen Milsom Wilke, Inc. (acoustics); SWA Group (landscape)

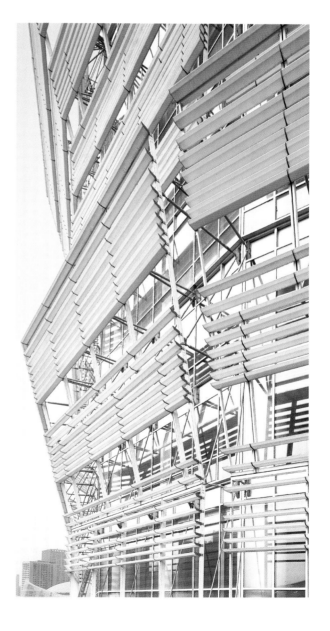

Opposite: Overall view of tower from the north

Top Left: Section diagram showing heliostat and hotel atrium

Bottom Left: View of tower in context at the lake's edge

Right: Close-up view of the canted aluminum screen shading system

103

Nominees Best Tall Building Asia & Australasia

Alamanda Office Tower
Jakarta, Indonesia

Completion Date: March 2013
Height: 131 m (430 ft)
Stories: 30
Area: 36,427 sq m (392,097 sq ft)
Use: Office
Owner/Developer: PT. Karyadeka Graha Lestari
Architect: DYXY Architecture + Interiors (design); PDW Architects (architect of record)
Structural Engineer: PT. Kinematika
MEP Engineer: PT. Malmass Mitra Teknik
Project Manager: PT. Arkonin
Main Contractor: PT. Tata Mulia Nusantara

ARK Hills Sengokuyama Mori Tower
Tokyo, Japan

Completion Date: August 2012
Height: 199 m (653 ft)
Stories: 47
Area: 136,100 sq m (1,464,968 sq ft)
Use: Office/Residential
Owner: Union Urban Redevelopment of Toranomon-Roppongi Area
Developer: Mori Building Co., Ltd.
Architect: Mori Building Co., Ltd. (design); Irie Miyake Architects & Engineers (architect of record); Pelli Clarke Pelli Architects (architect of record)
Structural Engineer: Obayashi Corporation; Yamashita Sekkei, Inc.
MEP Engineer: Kenchiku Setsubi Sekkei Kenkyusho
Main Contractor: Obayashi Corporation

City Tower Kobe Sannomiya
Kobe, Japan

Completion Date: February 2013
Height: 190 m (623 ft)
Stories: 54
Area: 73,500 sq m (791,147 sq ft)
Use: Residential/Hotel
Owner/Developer: Asahi-Dori 4-Chome Area Urban Redevelopment Union
Architect: Research Institute for Environmental Redevelopment (design); Tokyu Architects & Engineers, Inc. (architect of record)
Structural Engineer: Orimoto Structural Engineers
MEP Engineer: Tokyu Architects & Engineers, Inc.
Main Contractor: Obayashi Corporation

Dolphin Plaza
Hanoi, Vietnam

Completion Date: June 2012
Height: 134 m (441 ft)
Stories: 30
Area: 83,000 sq m (893,405 sq ft)
Use: Residential
Owner: TID Joint Stock Company; PVFC Land
Developer: TID Joint Stock Company
Architect: DP Architects Pte., Ltd. (design); Bureau Veritas (architect of record)
Structural Engineer: Consultancy Company of University of Civil Engineering
MEP Engineer: Chomthai Design and Consultants Co., Ltd.
Main Contractor: Hanoi Construction Joint Stock Company No. 1

Nominees Best Tall Building Asia & Australasia

Huarun Tower
Chengdu, China

Completion Date: July 2012
Height: 201 m (659 ft)
Stories: 39
Area: 72,730 sq m (782,859 sq ft)
Use: Office
Owner/Developer: China Resources Land (Chengdu) Ltd.
Architect: Callison, LLC
Structural Engineer: Guangzhou Rongbaisheng Structural Design Firm
MEP Engineer: PBA
Project Manager: Jizhun Fangzhong Architectural Design Associates
Main Contractor: CR Construction
Other Consultants: AECOM (landscape); Arup (façade); Lighting Design Alliance (lighting); Oculus Light Studio (lighting)

Net Metropolis
Manila, Philippines

Completion Date: October 2012
Height: 127 m (417 ft)
Stories: 26
Area: 43,200 sq m (465,000 sq ft)
Use: Office
Owner: The Net Group
Developer: 20–12 Property Holdings, Inc.
Architect: Oppenheim Architecture + Design (design); Leandro V. Locsin Partners (architect of record)
Structural Engineer: LSD and Associates
MEP Engineer: Meinhardt
Project Manager: JCLI International
Main Contractor: Monolith Construction and Development Corporation
Other Consultants: C|S Design Consultancy (interiors)

Pyne
Bangkok, Thailand

Completion Date: October 2012
Height: 161 m (528 ft)
Stories: 44
Area: 30,197 sq m (325,038 sq ft)
Use: Residential
Owner/Developer: Sansiri Public Company Limited
Architect: P & T Group
Structural Engineer: Infra Technology Sevice Co., Ltd.
MEP Engineer: P & T Group
Project Manager: SEA Consult Engineering Co., Ltd.
Main Contractor: Siphya Construction Co., Ltd.
Other Consultants: Diu Design Identity Unit Co., Ltd. (interiors); T.R.O.P., Ltd. (landscape)

Reflection Jomtien Beach
Pattaya, Thailand

Completion Date: May 2013
Height: Tower A: 234 m (768 ft); Tower B: 183 m (600 ft)
Stories: Tower A: 57; Tower B: 44
Area: 59,622 sq m (641,766 sq ft)
Use: Residential
Owner/Developer: Major Development
Architect: P & T Group
Structural Engineer: P & T Group
MEP Engineer: P & T Group
Project Manager: Stonehenge Inter Co., Ltd.
Main Contractor: Syntec Construction Public Co., Ltd.
Other Consultants: August Design Consultant Co., Ltd. (interiors); P Landscape Co., Ltd. (landscape)

Nominees Best Tall Building Asia & Australasia

Shenzhen Kerry Plaza Phase II
Shenzhen, China

Completion Date: March 2012
Height: 200 m (656 ft)
Stories: 41
Area: 79,000 sq m (850,349 sq ft)
Use: Office
Owner/Developer: Kerry Development (Shenzhen) Co., Ltd.
Architect: Skidmore, Owings & Merrill LLP (design); Shenzhen General Institute of Architectural Design and Research Co., Ltd. (architect of record)
Structural Engineer: AECOM
MEP Engineer: AECOM
Main Contractor: The First Construction Co., Ltd. of China Construction Third Engineering Bureau Corp., Ltd.
Other Consultants: Arup (façade); BC&A (interiors)

Yixing Dongjiu
Yixing, China

Completion Date: December 2012
Height: 244 m (800 ft)
Stories: 56
Area: 161,725 sq m (1,740,793 sq ft)
Use: Office/Hotel
Owner/Developer: Yixin Jinyuan Real Estate Co., Ltd.
Architect: ECADI
Structural Engineer: ECADI
MEP Engineer: ECADI
Main Contractor: Zhejiang Haitian Construction Group

Yokohama Mitsui Building
Yokohama, Japan

Completion Date: February 2012
Height: 152 m (499 ft)
Stories: 31
Area: 90,356 sq m (972,584 sq ft)
Use: Office
Owner/Developer: Mitsui Fudosan Co., Ltd.
Architect: Nikken Sekkei, Ltd.
Structural Engineer: Nikken Sekkei, Ltd.
MEP Engineer: Nikken Sekkei, Ltd.
Main Contractor: Taisei Corporation

Best Tall Building
Europe

Winner
Best Tall Building Europe

The Shard
London, United Kingdom

Completion Date: February 2013
Height: 306 m (1,004 ft)
Stories: 73
Area: 111,000 sq m (1,194,790 sq ft)
Use: Residential/Hotel/Office
Owner: London Bridge Quarter, Ltd.
Developer: Sellar Property Group
Architect: Renzo Piano Building Workshop (design); Adamson Associates Architects (architect of record)
Structural Engineer: WSP Group
MEP Engineer: Arup
Project Manager: Turner & Townsend
Main Contractor: Mace

"The skyline of London has been redrawn by this building. It has awoken the enthusiasm for European architectural innovation."

Nengjun Luo, Juror, CITIC Heye Investment

At the time of its completion, The Shard became the tallest building in Western Europe. The iconic tower has redefined the London skyline and is already an international symbol for London. A mixed-use "vertical city," it offers more than 55,000 square meters of office space on 25 floors, three floors of restaurants, a 17-story hotel, 13 floors of apartments and a triple-height viewing gallery, as well as an open-air viewing floor on level 72. It is crowned with a steel-framed pinnacle and clad with shards of glass designed to blend into the sky. Standing next to London Bridge Station, one of London's busiest transport hubs, at the heart of London Bridge Quarter, London's newest commercial quarter, The Shard is a key part of the regeneration of London's South Bank.

The name "The Shard" is derived from the architect's description of the development as a "shard of glass" during planning stages. Its design was influenced by the irregular nature of the site.

The tapered form of the building provides efficient and economic floor design, with optimally sized floor plates conducive to its function as a multi-use development. Thus, offices on the lower floor make use of large open-plan spaces with minimal structural intrusion, while the upper floors suit the uses of the hotel rooms and apartments, which require smaller floor plates. Moving further up into the spire, steel beams and columns with elegantly detailed connections help create an aesthetic, open space for the public to appreciate the views.

Matching the structure to the different uses allowed efficient use of materials, reducing both cost and the amount of embodied carbon on the project. It also maximized the net lettable area for the client.

Delivering Europe's tallest tower in record time drove structural engineers and contractors to rethink the basic principles of construction and use new techniques to go higher and faster than had been conducted in the UK previously.

To overcome the challenges of building a skyscraper safely in central London, adjacent to a major transport hub, the team delivered a number of firsts: the first core to be built by top-down construction, the UK's largest concrete pour, the first use of jump-lift construction, the first inclined hoist in the world, and the first crane to be supported on a slipform. A specially designed laser-guided drilling rig was used to surgically place pilings among Victorian-era underground utilities and ancient archaeological finds. Top-down construction allowed the first 23 stories of the concrete core and much of the surrounding tower to be built before the basement had been fully excavated. This technique was a world first and saved four months on the complex program.

The distinct tapering form is achieved in five structural parts. From basement level three to the 72nd floor there is a reinforced concrete core. The first 40 floors are a composite steel frame, while a post-tension concrete frame runs up to level 60, with a traditional reinforced concrete frame taking the project to level 72. The spire,

Previous Spread
Left: Overall view of tower
Right: Base of tower

Current Spread
Opposite: Night view from the east
Top Left: Public observatory
Bottom Left: Typical office floor

Below: Floor plans – typical residential at level 39 (top) and typical office at level 32 (bottom)

Jury Statement

The developers of The Shard showed remarkable tenacity in bringing it to fruition. The level of determination to wring economic success and poetics out of the project while still supporting public life at street level was remarkable. Through more than a decade of design revisions and public inquiries, the project team was unwavering in its determination to do more than impose a tall building on a neglected but architecturally rich neighborhood. Their determination was to secure the future of the London Bridge Quarter district itself.

The building is both heroic on the skyline and beautifully executed at the scale of the pedestrian, and clearly prioritizes public transportation over the automobile. The Shard is an amazing accomplishment – few tall buildings in historic city centers have been executed as successfully. It's rare that a building of this size so clearly reveals itself as the outcome of thoughtful consideration of the future and an abiding respect for its historic surroundings.

Left: Elevation sketch
Opposite: Aerial view of tower

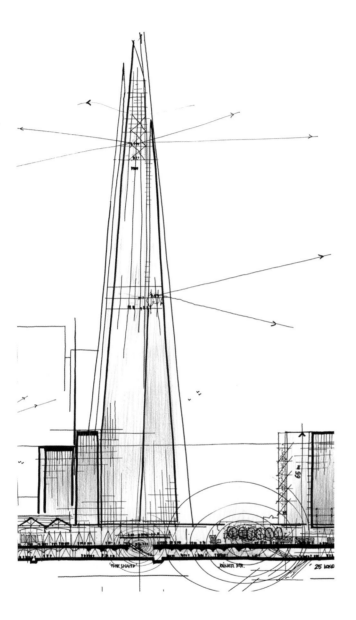

to level 87, is made of steel. The mixture of concrete and steel increased the efficiency of the structure. The design negated the need for expensive tuned mass dampers by building the hotel and apartment levels in concrete, sandwiched between the steel office floors and spire.

Post-tensioned concrete was more suitable for the smaller spans higher up the building and saved 550 millimeters per floor. The concrete also provided structural damping, thus saving money and weight and releasing a further two floors as lettable space. In addition, fabricated uniform-depth steel beams acting compositely with the concrete floor slabs optimized the space in the ceilings for services.

The Shard is intended to regenerate and energize South London. The development promotes sustainable travel by including only 48 car parking spaces and features a major refurbishment of the adjacent London Bridge station, which handles 54 million passengers a day.

"We often forget the hard work it takes to actually construct a tall building in densely populated cities. To build such an iconic, high quality building, safely and quickly in the middle of the City of London, is truly remarkable."

Robert Okpala, Juror, Buro Happold

Finalist
Best Tall Building Europe

ADAC Headquarters
Munich, Germany

Completion Date: March 2012
Height: 93 m (305 ft)
Stories: 23
Area: 125,000 sq m (1,345,489 sq ft)
Use: Office
Owner: ADAC
Architect: Sauerbruch Hutton
Structural Engineer: Werner Sobek Group (design); henke + rapolder Ingenieurgesellschaft (peer review)
MEP Engineer: NEK Beratende Ingenieure
Main Contractor: ARGE Neubau; ADAC Zentrale
Other Consultants: Transsolar (energy concept); Werner Sobek Group (façade)

"Sauerbruch and Hutton have become intuitive masters of their craft. A subtle and graceful massing and refined color palette settle the building comfortably into its Munich context."

Richard Cook, Juror, COOKFOX Architects

The new headquarters of the German Automobile Club (ADAC) features a dynamic design and harmonized concept, enriching the skyline. With 75,000 square meters of usable space above ground, and 50,000 square meters of usable space below ground, the building offers all 2,400 ADAC Munich staff an abundant amount of room, including offices, a large conference and training center, a restaurant, and a printing plant.

The building complex consists of a five-story plinth building, upon which a high-rise tower is placed. The ground plan of the star-shaped plinth building measures 187 meters in a longitudinal direction and 107 meters in a transverse direction, while the high-rise building atop the plinth measures 35 by 65 meters. The high-rise tower was placed next to the railway tracks so that it neither cast shadows onto the courtyard and neighboring buildings, nor dominated the street.

A large number of demanding engineering solutions were required to support the high-rise tower, to build

Finalist
Best Tall Building Europe

New Babylon
The Hague, Netherlands

Completion Date: January 2013
Height: City Tower: 142 m (465 ft); Park Tower: 102 m (333 ft)
Stories: City Tower: 45; Park Tower: 31
Area: 143,500 sq m (1,544,615 sq ft)
Use: Residential
Owner/Developer: Babylon Den Haag BV (collaboration between SNS Property Finance BV and Fortress BV)
Architect: MVSA Architects
Structural Engineer: Corsmit Raadgevende Ingenieurs
MEP Engineer: bv Adviesburo T&H
Main Contractor: Bouwcombinatie New Babylon vof; Ballast Nedam NV en Boele & Van Eesteren

"Reusing the old Babylon, while increasing density and providing a new architectural statement, was not an easy proposition, but they have managed to achieve all of this and more."

Robert Okpala, Juror, Buro Happold

New Babylon transforms an introverted 1970s building into a 21st-century, high-rise cityscape, integrating prime work and living space and connecting occupants with an ever-growing international community. The Babylon renovation project began in 2003 with a brief to extend the existing Babylon complex at The Hague's central railway station, as part of a broader scheme to revitalize the center of The Hague.

To take maximum advantage of the building's structural integrity while avoiding costly and undesirable demolition work, the designers incorporated the existing structure within a two-tower high-rise complex. To meet the spatial requirements of New Babylon, the usable floor space had to be tripled to 143,500 square meters, while the footprint was to increase from 7,100 square meters to 10,500 square meters.

The towers were strategically designed to take advantage of the surrounding views, while fully conforming to local height restrictions. Visible from every corner of

Previous Spread
Left: Overall view of complex
Right: Roof gardens between original and new structures

Current Spread
Left: Looking down at sky gardens and connections between original and new structures
Opposite: Side by side comparison of original 1970s structures before and after renovation

> "The seamless integration of a dilapidated 1970s building into an inspiring modern development brings light and livability into this urban location."
>
> David Scott, Juror, Laing O'Rourke

The Hague, New Babylon is not only easily seen, but also easily reached by both private and public transport. Effectively, New Babylon is a "recycled building" that promotes new standards for urban development. When initially designed in the 1970s, the building was modeled on a series of stacked volumes, representing the multiple functions existing in the ancient city of Babylon. This was considered cutting edge, but by 2000, the brown interior and introverted design was out of sync with the times. New Babylon advances this concept to create new layers of form and function, creating new space for living and working in a busy city center.

The first two levels of New Babylon accommodate shops, restaurants, and other commercial premises. Wide, light-filled passageways accessible from each of the building's four sides guide the public through the complex. Inside the building the office floors are spatially interwoven with the lower shopping levels. The upper levels house offices and a conference center that can be reached via the sky lobby on the atrium's second level. The atrium roof makes the height of the two residential towers visible from inside the building. Several gardens on the plinth roof complement the vertical gardens in the shopping center.

The Park Tower on the northeast side and the City Tower on the southwest side contain 335 owner-occupied and rental apartments. Featuring street-level entrance halls and a concierge service, each tower provides easy access to the atrium and shopping center. The apartments start at level two and range from 90 to 280 square meters in area; each apartment has its own outdoor space in the form of a balcony or roof terrace.

Given Old Babylon's somewhat dark and sterile appearance and imposing scale, the designers wanted to introduce a human factor. To achieve this, a new, graduated façade was designed, with each step accentuating the function of the corresponding floor. An open retail plinth merges into alternating rows of offices and apartments, whose balconies blend smoothly

into the composition. By demolishing the above-ground car park and replacing it with underground parking, the redevelopment has created much-needed public space that complements the addition of living space to the complex. The open character of the façades enhances the cohesion between the plinth, with its shops and arcades, and the adjoining public spaces. Thanks to these characteristics, New Babylon integrates seamlessly with the city's spatial design, and enhances the dynamism and vitality of its immediate surroundings.

Jury Statement

This project takes the extraordinary step of converting a high-concept but dated design into something wholly different, subsuming the original into a larger, more ambitious but more contextually appropriate creation. Taking adaptive reuse to an extreme rarely seen, particularly with late International-Style buildings, New Babylon will play a critical role in the regeneration of its immediate surroundings. In a space-constrained urban future, this will be a reference project for the next generation.

Finalist
Best Tall Building Europe

Tour Total
Berlin, Germany

Completion Date: September 2012
Height: 69 m (226 ft)
Stories: 18
Area: 28,000 sq m (301,389 sq ft)
Use: Office
Owner/Developer: CA Immo Deutschland GmbH
Architect: Barkow Leibinger
Structural Engineer: GuD Planungsgesellschaft für Ingenieurbau mbH
MEP Engineer: Fürstenau & Partner Ingenieurgesellschaft mbH
Main Contractor: omniCon Gesellschaft für innovatives Bauen mbH
Other Consultants: BBM Müller-BBM (acoustics); Drees & Sommer (sustainability); energydesign Braunschweig GmbH (energy concept); hhp Berlin Ingenieure für Brandschutz (fire); Priedemann Fassadenberatung GmbH (façade)

"Tour Total shows us that it is possible, and indeed necessary, to expand the material palette of the high-rise beyond glass alone, demonstrating both the thermal and visual benefits of doing so."

Jeanne Gang, Jury Chair, Studio Gang Architects

Set in Europacity, a master plan for a new urban district of 40 hectares directly to the north of the main train station in Berlin, Tour Total is the headquarters for the French energy company Total and is the first building in this plan. It is a freestanding high-rise that gives the company and its 500 employees a clear identity for their headquarters in Germany. The DGNB (German Sustainable Building) Silver Certificate was an early goal and guided the design. Much of this was achieved through an intelligent façade system and energy reuse.

The freestanding tower defines a pedestrian passage on the west that leads to a new public space with restaurants and other amenities, located between the new tower and a planned adjacent urban block. The building acts as a pilot project for Europacity, and serves as an example, setting the tone aesthetically and sustainably for future construction within the plan. Tour Total demonstrates how a master plan can adapt and change to the specific requirements of individual clients.

Jury Statement

Tour Total establishes a strong anchor for a new urban district in the heart of Berlin. It achieves a remarkable balance between economics and sustainable performance; formal and aesthetic ambitions are integrated with functional and performative ambitions. The scale and elegance of the precast façade establishes visual interest along every face, while each face reacts individually to its surroundings. The addition of pre-assembly adds a layer of practical constructability that should inspire future designs in the district to aim high.

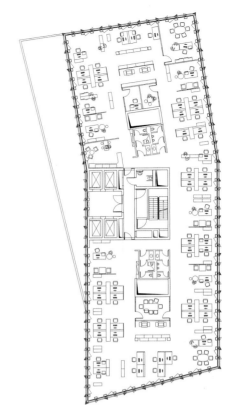

While conforming to the intention of the master plan, it also provokes and pushes back against it.

Tour Total combines a Berlin typology — the "raster façade" or façade-grid — with ambitions for sustainable energy use, which was a primary goal for the tenant. This is consistent with an idea of combining formal/aesthetic ambitions for the building with performative/functional ambitions, wherein both are integrated. Sustainability is not a project "add-on."

The raster façade is a sculpturally folded curtain, which wraps the volume, giving the building a dynamic effect in changing conditions of light and shadow. Historic Berlin façades typically operate with a material depth by which a façade may appear opaque from an oblique view and transparent from the front. Tour Total subverts this typology by creating a rippling visual effect through the precast façade elements. This effect is pronounced by oblique natural daylight and by electric lighting at night. For more information on the innovative façade system used in Tour Total see the Innovation Awards section where the system was awarded Finalist status, page 176.

The volume of the building generates well-lit and ventilated office floors. Every second window is operable to a safe opening distance, allowing natural ventilation. The form of the building reacts to a number of existing urban conditions. Its front is oriented to Heidestrasse and to the planned future park to the north. The overall form then folds, creating concave and convex sides, reacting to both the orthogonal edge of the Heidestrasse and to the radial system generated by the curving Minna-Cauer Strasse. A two-story arcade, defined by columns, wraps the building base, with closed and open arcades for the main entrance and a pedestrian path to the north. The arcade acts as a filter between the lobby and the exterior, and as a scaling device for the overall building.

Total was determined to express the ambition of sustainability and to achieve an appealing office space. The building is configured to provide flexible workspaces of a floor-plate depth that can be both naturally lit and ventilated. The building as a freestanding tower offers complex views and orientations from all sides. The ground floor consists of an entrance lobby and bistro opening out to the pedestrian passage generated by the scheme for usable exterior space, opening the building up to the city and emerging community.

Previous Spread

Left: Overall view from the south

Right: Close-up view of the faceted concrete façade panels

Current Spread

Opposite: Typical floor plan

Top: Typical upper floor office interior

Bottom Left: Overall view in context

Bottom Right: Lobby

"As the first building completed in the Europacity master plan, Tour Total sets a firm direction in terms of aesthetics and materials for the larger development to come."

Karen Weigert, Juror, Chicago Chief Sustainability Officer

Nominee
Best Tall Building Europe

Mercury City
Moscow, Russia

The architectural solution for Mercury City Tower was influenced by several factors, including a requirement for solar energy capture and the close proximity of considerably sized neighbors. This accounts for the knifelike profile of the tower, emphasized by the interrupted zigzag of the reinforced concrete structures at the façade. At the same time the traditional three-part massing of the volume, comprising a base, core, and crown, creates the impression of strength, reliability, and serenity. The designers gave priority to the integrity and completeness of the tower volume, together with the north façade which faces the main, northern entrance to Moscow-City, located on the Third Transport Ring Road.

The internal volume of Mercury City is planned so as to use the space most effectively. Apartments are located on levels 43 through 75 and have panoramic windows. Flexibility in the residential plan allows for the option of merging multiple units into one single larger unit.

Completion Date: July 2013
Height: 339 m (1,112 ft)
Stories: 75
Area: 173,960 sq m (1,872,490 sq ft)
Use: Residential/Office
Owner: CJSC Mercury City Tower
Developer: Mercury Development
Architect: Frank Williams & Associates; M.M. Posokhin
Structural Engineer: International High-Rise Construction Centre, LLC
MEP Engineer: International High-Rise Construction Centre, LLC
Main Contractor: HSG Zander

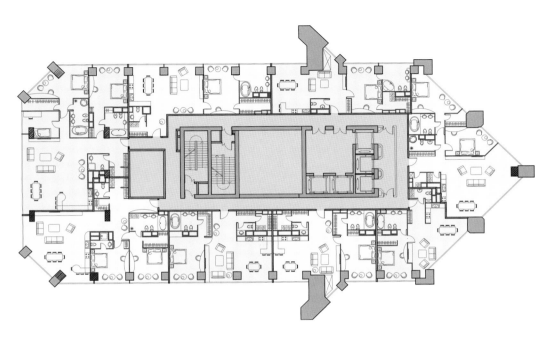

Opposite: Overall view of tower
Top: Typical residential floor plan
Bottom Left: Typical residential interior
Bottom Right: Close-up view of the façade

Nominee
Best Tall Building Europe

Unicredit Tower
Milan, Italy

The Unicredit headquarters is a complex of three towers, comprising the largest components of Porta Nuova Garibaldi, a seven-hectare, mixed-use development north of Milan's city center, which redevelops the abandoned rail yards adjacent to Stazione Garibaldi, forming a new gateway to the city.

Spiraling upward, the asymmetrical main tower culminates in a sculptural, stainless steel spire. Like the two smaller towers, the building is clad in reflective glass. Their narrow, curved forms enclose Piazza Gae Aulenti, a new public space. Facing the piazza, the façades incorporate sunshades, emphasizing the buildings' fluid shape. At the street level, the towers are clad in stone. Around the piazza, a ring-shaped canopy connects the podiums of the three towers. Two levels of shops are above the piazza, with additional retail and dining at the sunken level. The combined podium contains parking and a direct connection to the Stazione Garibaldi rail station.

Completion Date: December 2012
Height: 218 m (714 ft)
Stories: 33
Area: 26,708 sq m (287,483 sq ft)
Use: Office
Owner/Developer: Hines
Architect: Pelli Clarke Pelli Architects (design); Adamson Associates Architects (architect of record)
Structural Engineer: MSC Associati, S.r.l.
MEP Engineer: Ariatta Ingengneria dei Sistemi, S.r.l.; Buro Happold
Main Contractor: Colombo Costruzioni, S.p.A.
Other Consultants: J&A Consultants S.n.c. (quantity surveyor); Jan Gehl Architects (urban planning); Studio Leonardo Corbo (life safety)

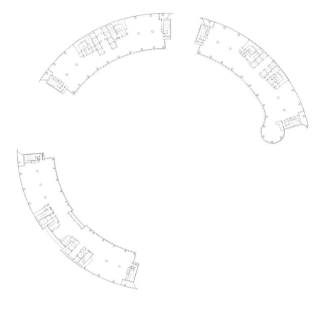

Opposite: Overall view of complex

Top Left: Overall view of the main tower with spire

Bottom left: Public plaza at the base of the main tower

Top Right: Typical floor plan

Bottom Right: Public plaza at the center of the complex

Nominees Best Tall Building Europe

No. 1 Great Marlborough Street
Manchester, United Kingdom

Completion Date: September 2012
Height: 106 m (348 ft)
Stories: 37
Area: 14,975 sq m (161,190 sq ft)
Use: Residential (student housing)
Owner/Developer: Student Castle
Architect: Hodder+Partners
Structural Engineer: WSP Group
MEP Engineer: Brentwood Group
Project Manager: Canmoor
Main Contractor: Shepherd Construction
Other Consultants: Cladtech (façade); Sharpes Redmore (acoustics)

Torre Unipol
Bologna, Italy

Completion Date: December 2012
Height: 125 m (410 ft)
Stories: 28
Area: 29,000 sq m (312,153 sq ft)
Use: Office
Owner/Developer: Unifimm
Architect: Open Project
Structural Engineer: Majowiecki
MEP Engineer: Betaprogetti
Main Contractor: Cooperativa Muratori e Braccianti di Carpi; Tosoni

Varyap Meridian Block A
Istanbul, Turkey

Completion Date: December 2012
Height: 188 m (618 ft)
Stories: 52
Area: 5,100 sq m (54,896 sq ft)
Use: Residential/Office
Owner/Developer: Varyap
Architect: RMJM (design); Dome Mimarlik (architect of record)
Structural Engineer: Teknik Yapi; Buro Happold
MEP Engineer: Buro Happold
Main Contractor: Varyap
Other Consultants: Cwg Danismanlik, Ltd. (façade); Tuncay Akdag (fire)

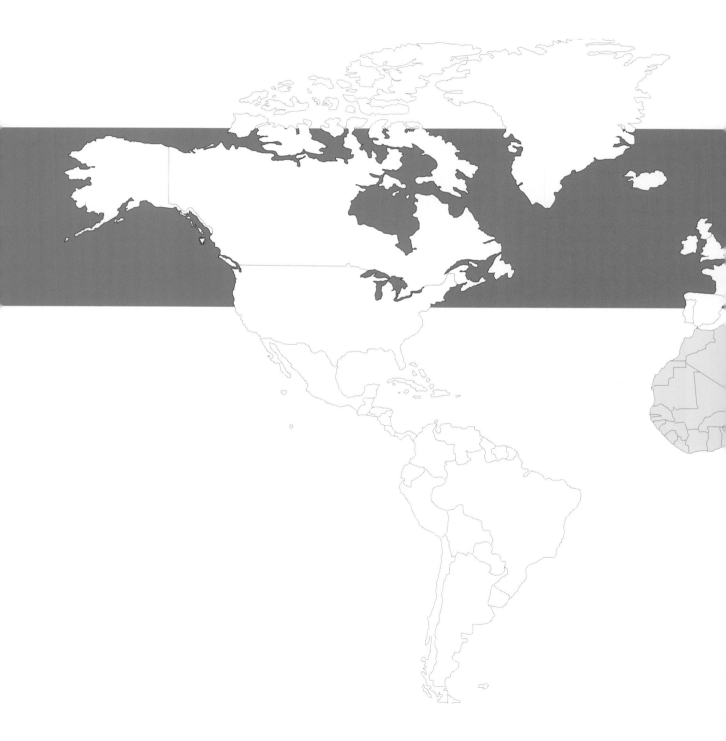

Best Tall Building
Middle East & Africa

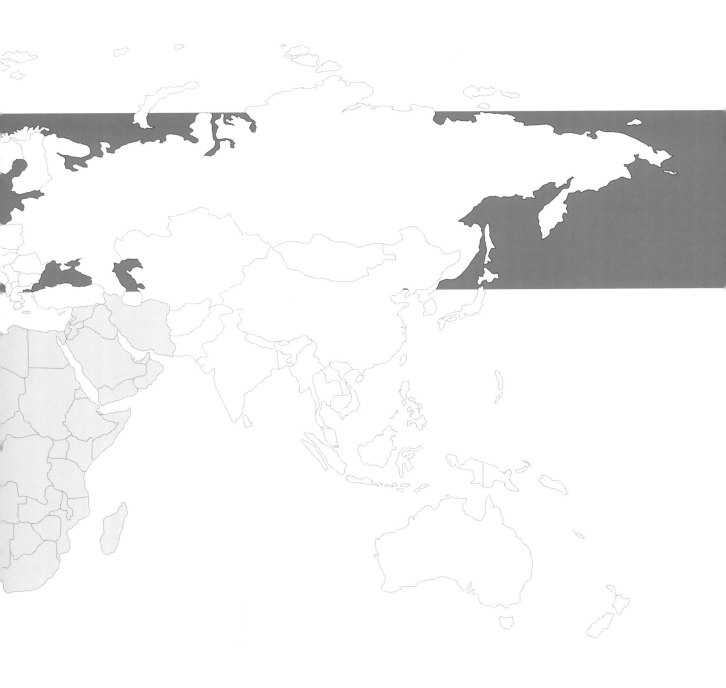

Winner
Best Tall Building Middle East & Africa

Sowwah Square
Abu Dhabi, United Arab Emirates

Completion Date: December 2012
Height: Al Khatem, Al Maqam: 155 m (509 ft); Al Sarab, Al Sila: 131 m (430 ft)
Stories: Al Khatem, Al Maqam: 37; Al Sarab, Al Sila: 31
Area: 529,400 sq m (5,698,414 sq ft)
Use: Office
Owner/Developer: Mubadala Real Estate & Infrastructure
Architect: Goettsch Partners (design); Serex International (architect of record)
Structural Engineer: Oger International (design); Serex International (engineer of record); Thornton Tomasetti (peer review)
MEP Engineer: Oger International (design); Environmental Systems Design, Inc. (peer review)
Main Contractor: Oger Abu Dhabi
Other Consultants: Hann Tucker Associates (acoustics); Integrated Environmental Solutions, Ltd. (LEED); Jenkins & Huntington, Inc. (vertical transportation); Martha Schwartz Partners, Ltd. (landscape); MFD Security Ltd. (security); One Lux Studio, LLC (lighting); RWDI (wind); WSP Group (fire)

"This is a very powerful composition where sustainable features have been embedded from the design stage as part of the core functions within the complex."

Karen Weigert, Juror, Chicago Chief Sustainability Officer

Sowwah Square is a major new commercial development on Abu Dhabi's Al Maryah Island. The city's new urban framework plan, entitled Plan Abu Dhabi 2030, has designated the previously undeveloped island and the adjacent edges of Mina Zayed and Reem Island as the city's new Central Business District. The project totals over 290,000 square meters of office space and features the iconic new headquarters building for the Abu Dhabi Securities Exchange, surrounded by four office towers, all overlooking the water. In addition, the project integrates two levels of retail and two parking structures. A generous landscaped plaza connects the four buildings and the exchange.

The stock exchange building at the center of the complex is an iconic, four-level facility. Glass-enclosed with a roof the size of a football field, the building rises 27 meters above a 49-meter-diameter water feature on massive stone piers. The four granite piers house the stairs, mechanical risers, and service elements for the exchange. The four office towers frame the stock

exchange building. The first full office floor of each building starts 34 meters above the ground level, providing a transparent, open lobby and elevating the views from tenant floors.

Beneath the plaza, a two-story retail podium weaves through the development, providing 23,220 square meters of upscale shopping along the waterfront. At the north and south boundaries of the site, two parking structures, partially submerged, serve the complex with more than 4,800 parking spaces.

Sowwah Square is the first mixed-use project in Abu Dhabi to be pre-certified LEED-CS Gold based on sustainable initiatives. However, the complex looks beyond the LEED certification process to emphasize a sustainable design approach throughout, integrating both active and passive sustainable design strategies.

The environmentally responsive enclosure system uses a mechanically ventilated cavity and a double-skin façade system over large portions of the office buildings. These elements mitigate the 40 °F interior/exterior temperature differential and protect building occupants from the intense sandstorms and the constant corrosive mist of the neighboring Gulf coast.

The double-skin cavities run uninterrupted along the entire height of the four office towers, starting from the fourth floor and extending to the penthouse mechanical floors. Within these cavities, active solar shades continuously track and adjust for the sun angle to provide optimal shading to the building's interior. The cavity is sealed to protect the gears from airborne particulates.

Active solar shading and glass selection keep the cavity from increasing the internal radiant temperature. To minimize the amount of solar energy penetrating the outer layer of the double-skin system, an outboard fin with a very high shading coefficient (76 percent) was selected. The remaining energy was then blocked from reaching the inner façade by the active shading; however,

Previous Spread

Left: Base of tower with dramatic 34-meter height before first full floor

Right: View from public plaza surrounding the towers

Current Spread

Opposite: Overall view of complex at night

Right: Detail section through double-skin façade (top), diagram showing location of double-skin façades (bottom)

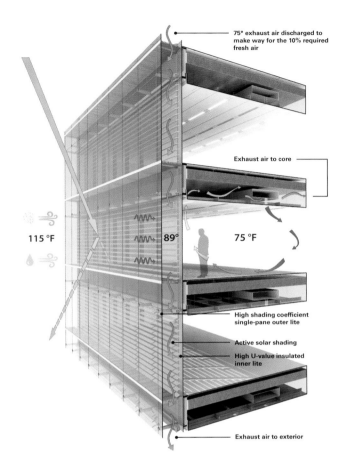

Jury Statement

Sowwah Square stands out as consciously sustainable and warmly inviting, yet a formally disciplined project in a region where achieving such aims have historically proven difficult. The interdependent elements work together such that the project functions as an integrated machine. From sun-tracking shading devices to elevated lobbies with views of cool roofs and the sweep of the harbor, little seems to have been left out of the calculations. That a building in this climate could support as much glass as it does is a testament to the possibilities of well-orchestrated design.

Its clear massing and organization helps to ensure that the design will not appear dated and will continue to support pedestrian activity, even as new sustainable technologies come into use. The elevation of the stock exchange on pedestals reinforces the importance of the commercial activity inside, while creating an enticing, shaded communal space beneath, lending a calming tone to the monumental surroundings. The curving approaches at the waterside promenade promise to soften the edges of the angled towers, while clearly pointing the way to the center of activity.

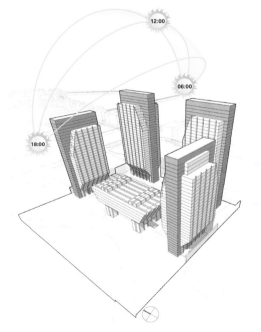

its presence contributed to an elevated air temperature within the double-skin cavity.

To alleviate the accelerated temperature and achieve the moderating air buffer, the warm cavity air needs to be flushed out using an air source cooler than the natural air temperature. The solution was to collect the exhaust air from the tower offices and, instead of allowing it to escape into the atmosphere, redirect it back down the double-skin cavities, where it is exhausted at the

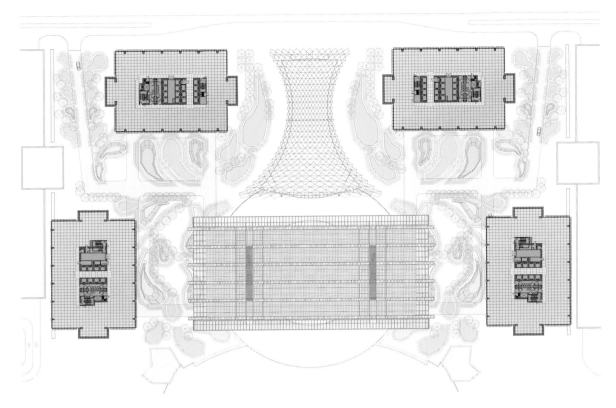

Top: Typical tower plans
Bottom: Tower base and cable-supported lobby enclosure
Opposite Top: Interconnecting plaza looking over the water
Opposite Bottom: Office tower lobby

fourth-floor mechanical level. Sensors within the cavities modulate dampers at the top of the building, directing the air to the optimal zones of the cavity depending on the time of day and outdoor temperature. Additional dampers will allow filtered exterior air to enter directly into the cavity during economizing periods such as night and winter, when the outdoor air is cooler than the collected exhaust air.

Through these efforts, the design team expects the double-skin cavity to be an average temperature of 89 °F (32 °C) when the exterior temperature reaches 115 °F (46 °C). This condition will allow the high U-value of the insulated inner glazing to more easily block the air cavity's radiating energy. Most importantly, calculations estimate that the double-skin system designed for Sowwah Square will generate savings of 7,200 kWh of electricity per day across all four towers and provide a more comfortable thermal environment near the perimeter wall, all while protecting itself from the harsh external elements.

"Despite the insanity of building single-skin, all-glass towers in the intense solar environment that is the Middle East desert, Sowwah Square is one of the first high-rise uses of a double-skin façade in that region."

Antony Wood, Juror, CTBUH

Finalist
Best Tall Building Middle East & Africa

6 Remez Tower
Tel Aviv, Israel

"The combination of types of glazing, operable panels, and vision panels – for the benefit of achieving solar shading – create an interesting façade pattern."

Jeanne Gang, Jury Chair, Studio Gang Architects

The 6 Remez tower is an unusual residential tower in several aspects. Perhaps most striking is its structural system, which uses an off-center core and minimal interior columns, a design move more common to office buildings, so as to afford wide, sweeping views to occupants and maximum flexibility for tenant improvements.

The core has two lobbies on each typical floor. The front lobby acts as a reception area, which can be considered an extension of the apartments themselves, while the second lobby is located in the back for service purposes. This concept enhances firefighter safety, by defining the back lobby as a safe deployment space on each floor.

Aesthetically, the tower reads as a complete object, sidestepping volumetric complexity in favor of simplicity. The composition is that of a monolithic urban sculpture, wrapped in a lace-like glass skin. The inner glass shades create a rhythm of gentle hues and

Completion Date: July 2013
Height: 118 m (387 ft)
Stories: 32
Area: 15,900 sq m (171,146 sq ft)
Use: Residential
Developer: CTU Investments, Ltd.
Architect: Moshe Tzur Architects and Town Planners, Ltd. (design); Do Architects (architect of record)
Structural Engineer: David Engineers, Ltd.
MEP Engineer: Iser Goldish Consulting, Ltd.; Bar Akiva Engineers, Ltd.
Project Manager: Waxman Govrin Geva Engineering, Ltd.
Main Contractor: U. Dori Group, Ltd.
Other Consultants: E.S.L–Eng.S.Lustig Consulting Engineers, Ltd. (vertical transportation); Pitsou Kedem Architect (interiors); S. Netanel Engineers & Consultants, Ltd. (life safety); TEMA – Urban Landscape Design (landscape)

textures, which account for the light and shimmering appearance of the tower. This is the first time that a tower façade has been entirely covered with an aluminum wire mesh, which also serves to conceal the VRV (Variable Refrigerant Volume) system located in each apartment on the eastern side of the tower.

A new curtain wall system was developed for this project. This system includes an inward-opening window integrated into a structural curtain wall system. This specific detail played a crucial part in shaping the overall appearance of the structural curtain wall, which also incorporates shades in the spandrels as well as in the vision glass. The low-E coating complements the shading system, while low-iron exterior glass affords thermal values appropriate to the hot climate. Electronically operated Venetian blinds complete the protective system. The effect of all these elements working together is to create a unique texture for the whole tower.

Located in the heart of Tel Aviv next to a main east-west urban axis, the tower is oriented west toward the sea; the tower's plan is asymmetrical, and serves to connect the tower to its urban and functional context. The balconies all face west, while smooth glass façades create a clean geometric volume on the north and south sides, beyond which the balconies do not extend. The aluminum wire mesh forms the finish material for the core exterior concrete walls, and is carefully articulated in two directions, creating a constantly changing appearance as the sun arcs overhead throughout the day.

The project includes a small park north of the tower. Plantings in the park are located strategically so as to minimize undesirable wind effects from the tower for park users. With a subtle entrance via the garden, rather than directly from the street, the park is open to the public at all times. Its landscape design is based upon the typology of a leaf in plan view, and features reflecting and biological pools. The arrangement of the basement, car park and overall building footprint leave 20 percent of the plot free of built structures or hardscapes.

Jury Statement

As Tel Aviv urbanizes and rapidly grows its skyline, it is refreshing to see a building that gives over so much of its site to public space. The tower is modest in presentation from without, with projecting glass fins sheltering its balconies, and is generous within, affording multi directional city views from each residence. By engineering the floor slabs to hang off the asymmetrical core without imposing massive view-blocking diagonals, the designers have added to the elegance and deceptive simplicity of the project.

Previous Spread
Left: Overall view of tower from the northwest
Right: Looking up façade of the tower

Current Spread
Right: Typical upper floor plan
Opposite Top Left: Base of tower with park in foreground
Opposite Bottom Left: View from one of the expansive balconies
Opposite Right: View of the balconies running up the west façade

"The jury appreciated the simple and elegant solution for this tower, which has a confidence in its straightforward massing against the rapidly changing Tel Aviv skyline."

Richard Cook, Juror, COOKFOX Architects

Finalist
Best Tall Building Middle East & Africa

Gate Towers
Abu Dhabi, United Arab Emirates

Completion Date: August 2013
Height: 238 m (781 ft)
Stories: 66
Area: 465,871 sq m (5,014,594 sq ft)
Use: Residential
Owner/Developer: Aldar Properties PJSC
Architect: Arquitectonica (design); Khatib and Alami (architect of record)
Structural Engineer: Khatib and Alami (design); Arup (peer review)
MEP Engineer: Khatib and Alami
Project Manager: Hill International, Ltd.
Main Contractor: Arabian Construction Company; Orascom
Other Consultants: Exova Warrington (fire); Meinhardt (façade); Pelton Marsh Kinsella (accoustics); Thomas Bell Wright International Consultants (façade)

"The introduction of the 'horizontal' – the potential urban plane – into the increasingly 'vertical' of our cities, as seen here at Gate Towers, is an urgent urban imperative."

Antony Wood, Juror, CTBUH

The Gate Towers consist of a series of towers that act as pillars beneath a curving lintel. This renders the effect of a monumental portal that defines the threshold to the Shams Abu Dhabi district, a newly created land mass formed as an extension of the Central Business District on exposed tidal sands within a fringe of mangroves at approximately five to seven meters above sea level. The area is gradually developing at a high density and growing in prominence. From a development standpoint, the concept was derived from two maxims. First, to make a statement of intent which is visible from the main part of the city. Second, to form a gateway to the remaining sites within the Shams Abu Dhabi precinct on Reem Island.

This project has a total of 3,533 luxury residential apartments in three towers connected at their top with a two-level skybridge structure, which contains 21 large luxury penthouses. This composition is adjoined by a 22-story, horseshoe-shaped building. The complex includes three swimming pools and four water features,

Previous Spread
Left: Overall view of complex
Right: Looking down through one of the skybridge openings

Current Spread
Left: Construction photo series showing installation of skybridge spans
Opposite Top: Looking up at the skybridge cantilever
Opposite Bottom: The u-shaped "Arc" building, the visible large openings house common sky gardens for the residents

"An ingenious combination of skybridge and building – the construction of which is an impressive engineering feat – fully embodies the harmony of public space and privacy."

Nengjun Luo, Juror, CITIC Heye Investment

with car parking in three subterranean levels. The podium includes two levels of retail shopping mall space.

The Capping Bridge across the towers is an extraordinary achievement, involving precise calculation and steel fabrication. The two 750-metric-ton spans between the towers were fabricated on the ground and vertically strand-jacked into position 238 meters up – the heaviest and highest lifts ever attempted for a real estate project. The cantilevered portion on the East Tower (Tower 5) was an act of precision using smaller truss modules installed by a traveling gantry above.

An innovative system of tension cables were used to provide an underside working platform for installing soffit panels and abseiling hooks. Tension cables were strung between two towers at three-meter centers, with periodic connections into the structure above, and nylon nets were stretched between them. This was the fastest way to achieve a platform 230 meters in the air.

The Arc (Tower 7) has a hanging garden feature, which brings small secluded pockets of landscape up to the residential levels of the building. There are 14 hanging gardens in total, each accessible from the adjacent corridor with its own irrigation system, fountains, and stepped seats.

Environmental effects were addressed in the design and were a guiding factor in the selection and installation of the mechanical and electrical equipment, which featured bipolar filters and pre-cooling of the fresh air handling units, which improved indoor air quality and movement while controlling the temperature throughout the building.

The cooling provisions for the building are provided by a centralized district cooling plant, which serves all the towers and podium, reducing energy consumption compared to conventional systems of chillers installed individually on each of the towers.

Jury Statement

The Gate Towers allude to a future in which tall buildings appear at such density that many of the amenities of urban habitat on the ground can and should be replicated at height. While clearly symbolic as an ageless gesture of welcome, the towers also afford a sophisticated experience for inhabitants. Moments of conviviality not normally experienced at height, such as hanging gardens, are cross-pollinated with experiences that could only come from pioneering engineering at height, such as spectacular skybridges.

Nominee
Best Tall Building Middle East & Africa

JW Marriott Marquis
Dubai, United Arab Emirates

The JW Marriott Marquis hotel is a twin-tower complex on an L-shaped plot in a prime location between Sheikh Zayed Road, the main road through Dubai, and Business Bay, a prominent business hub, adjoining the future Dubai Creek extension to the south. The towers currently hold the title for the tallest all-hotel function buildings in the world.

The hotel is designed in the expressionist style, inspired by the date palm, a symbol highly evocative of Arabian culture. Its detailing replicates the trunk of the palm. The protruding sections of the tower that give the building its textured appearance are not only aesthetic, playing with both light and shadow, but also translate to a diversity of room sizes, which increases versatility for the hotel operator. The two towers are located symmetrically on different axes in the far corners of the plot, to minimise overlooking each other and to maximise views of the promenade, Burj Khalifa, Safa Park and the Arabian Gulf.

Completion Date: November 2012
Height: 355 m (1,166 ft)
Stories: Tower 1, 2: 82
Area: 320,311 sq m (3,447,828 sq ft)
Use: Hotel
Owner/Developer: Emirates Airlines
Architect: Archgroup Consultants
Structural Engineer: BG&E; Archgroup Consultants
MEP Engineer: Ian Banham and Associates
Main Contractor: Brookfield Multiplex
Other Consultants: Fortune Consultants, Ltd. (vertical transportation); Kardorff (lighting); LWDesign (interiors); Meinhardt (façade); Verdaus (landscape)

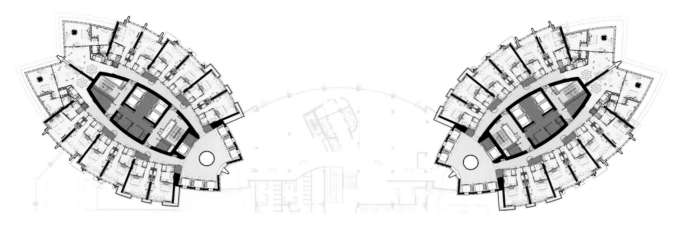

Opposite: Overall view from the business bay canal

Top: Typical floor plan

Above Left: Context view

Bottom Left: View of seven-story podium

Right: Close-up view of the faceted façade

Nominees Best Tall Building Middle East & Africa

Diplomat Commercial Office Tower
Manama, Bahrain

Completion Date: February 2012
Height: 158 m (517 ft)
Stories: 36
Area: 44,615 sq m (480,232 sq ft)
Use: Office
Owner/Developer: National Hotels Co.
Architect: Mohamed Salahuddin Consulting Engineering Bureau
Structural Engineer: Mohamed Salahuddin Consulting Engineering Bureau
MEP Engineer: Mohamed Salahuddin Consulting Engineering Bureau
Project Manager: Ismail Khonji Associates
Main Contractor: Chase Perdana Sdn Bhd

Faire Tower
Ramat-Gan, Israel

Completion Date: January 2013
Height: 100 m (326 ft)
Stories: 31
Area: 23,580 sq m (253,813 sq ft)
Use: Residential
Owner/Developer: Faire Fund
Architect: Canaan-Shenhav Architects
Structural Engineer: Israel David Engineers, Ltd.
Project Manager: Friedman-kamar Engineers, Ltd.
Main Contractor: A. Dori Group
Other Consultants: Leshem-Sheffer Environmental Quality, Ltd. (sustainability); Barr-Shoval Interior Design & Architecture (interiors); R.S. Cohen Safety Engineering, Ltd. (life safety)

Frishman 46
Tel Aviv, Israel

Completion Date: February 2013
Height: 107 m (351 ft)
Stories: 28
Area: 18,246 sq m (196,398 sq ft)
Use: Residential
Owner/Developer: EURO-SAT Investments, Ltd.
Architect: MYS Architects
Structural Engineer: Ben Avraham S. Engineers, Ltd.
MEP Engineer: Bar Akiva Engineers, Ltd.; S. Netanel Engineers & Consultants, Ltd.; Zvi Ronen Engineers, Ltd.
Project Manager: Waxman Govrin Geva Engineering, Ltd.
Main Contractor: U. Dori Group, Ltd.
Other Consultants: Alex Meitlis Architecture & Design (interiors); Zur Wolf Landscape Architects (landscape)

CTBUH 10 Year & Innovation Awards Criteria

10 Year Award

The CTBUH Best Tall Building awards, like most awards programs, recognize new buildings – based partly on the stated design intentions of these buildings. It is increasingly being recognized, however, that the industry needs to focus on actual "performance" rather than "best intentions" and thus, this year, the CTBUH has created a new "10 Year Award," which recognizes proven value and performance (across one or more of a wide range of criteria) over a period of time. This new award thus gives an opportunity to reflect back on buildings that have been completed and operational for at least a decade, and acknowledge those projects which have performed successfully long after the ribbon-cutting ceremonies have passed.

Evidence must be provided of performance in any category, including but not limited to: contribution to urban realm, contribution to culture/iconography, social issues, internal environment, occupant satisfaction, technical/engineering performance, environmental performance, energy performance, etc.

For a building to be considered for the CTBUH 10 Year Award, it must have been completed 10–12 years since the current award year. (e.g., for the 2013 award, a project must have a completion date between January 1, 2001 and December 31, 2003).

Innovation Award

This award recognizes a specific area of *recent* innovation in a tall building project that has been incorporated into the design, or implemented during construction, operation, or refurbishment. Unlike the CTBUH Best Tall Building awards, which consider each project holistically, this award is focused on one special area of innovation in the project – thus not the building overall. The areas of innovation can embrace any discipline, including but not limited to: technical breakthroughs, construction methods, design approaches, urban planning, building systems, façades, interior environment, etc.

The important criteria for judging is that the submission outlines succinctly the area of innovation, in comparison to standard benchmarks. The Innovation award can include recognition of a breakthrough that may not yet have been implemented in a specific building, but has been thoroughly tested.

The project must clearly demonstrate a specific area of innovation within the design and/or construction that is new and pushes the design of tall buildings to a higher level. The area of innovation should demonstrate an element of adaptability that would allow it to influence future tall building design, construction, or operation in a positive way.

Winner
CTBUH 10 Year Award

30 St Mary Axe
London, United Kingdom

"This tower proves that innovation can allow the office building to transcend perceived limitations and become a loved icon for its confidence within the context of a historic fabric."

Richard Cook, Juror, COOKFOX Architects

The inaugural winner of the CTBUH 10 Year Award, 30 St Mary Axe (The Gherkin), helped to define a modern, open, and progressive image for one of the world's oldest financial centers and set a benchmark in architectural quality for a new generation of tall buildings. The Gherkin has also been extraordinarily embraced by the public. In 2012, 3,000 people attended the Open City event to look inside, some queuing from 2am, with twice that number turned away.

As well as appearing on a first-class stamp, the tower has been used extensively in the promotion of London through advertising, notably as the symbol of London on Olympic bid posters. Thus, the Gherkin more than satisfies the awards criteria for "contribution to culture/iconography." However, the building is not only a cultural success, but a commercial one, consistently commanding higher rents than its peers in the City.

Under the engineering performance heading, the building's tapering form and diagonal bracing structure

Completion Date: 2003
Height: 180 m (590 ft)
Stories: 40
Area: 64,470 sq m (693,949 sq ft)
Use: Office
Owner: Evans Randall; IVG
Developer: Swiss Re
Architect: Foster + Partners
Structural Engineer: Arup
MEP Engineer: Hilson Moran Partnership Ltd.
Project Manager: RWG Associates
Main Contractor: Skanska AS
Other Consultants: BDSP Partnership Consulting Engineers (environmental); RWDI (wind)

have afforded numerous benefits that continue today: programmatic flexibility, naturally ventilated internal social spaces, and ample, protected public space at the ground level.

The Gherkin has performed exceptionally well in high winds – its robust aerodynamic form counteracts the movement that would otherwise be felt in a building of its height. Environmentally, this form, which slims toward the base and the apex, creates external pressure differentials that are exploited to drive a system of natural ventilation during the summer months, and enabled the creation of a generous, comfortable plaza at street level, which is protected from high winds by the tower's form.

The Gherkin's accommodating structure has had follow-on benefits in the internal environment and occupant satisfaction category. Column-free floor plates, and a fully glazed façade open the building to light and views. Six radial fingers of accommodation on each floor, with light wells between, combine the benefits of both curvilinear and rectilinear configurations, maximizing the proportion of naturally lit office space. Atria between the radiating fingers of each floor link vertically to form a series of informal breakout spaces that spiral up the building. As the occupancy has shifted from sole tenant to more than 14 firms, these "green lungs" have continued to provide valuable internal social space within the dense medieval street pattern of the City of London.

The geometry of the tower demanded an innovative system for the fabrication of individual cladding panels, due to the high level of variation. The 3D computer model of the system was linked directly to the production line, with major implications for the subsequent construction of complex buildings around the world.

The design placed a high priority on flexibility. Every possible configuration within the building, from cellular

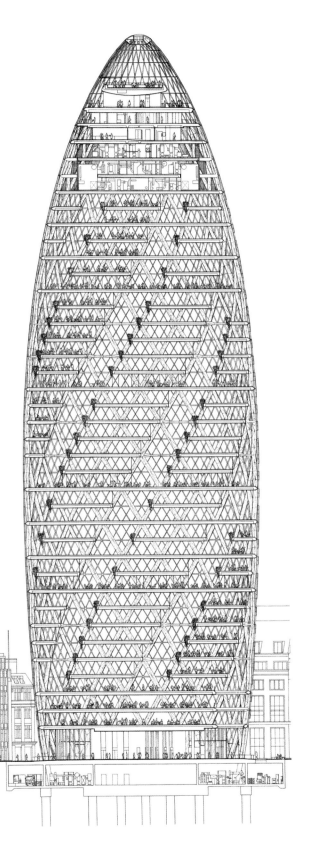

Previous Spread

Left: Aerial view showing 30 St Mary Axe within the dense medieval streets of London

Right: View down one of the atria separating the six "fingers" of the floor plan

Current Spread

Opposite: Building entrance

Left: Building section

Below: Detail section showing the ventilation strategy during the summer months

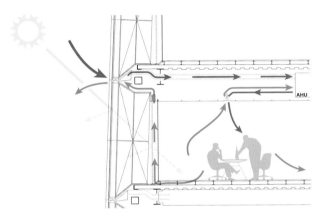

Jury Statement

Rarely does a single building have such a profound impact on the history of the high-rise as 30 St Mary Axe – The Gherkin. In just a few short years, the Gherkin not only launched the trend of affectionately naming London skyscrapers, it paved the way for the current generation of non-orthogonal tall buildings that now have become a quintessential feature of the city. The Gherkin showed us not only that skyscrapers could be more than the simple upward extrusion of a floor plan; it showed us that we could demand more from skyscrapers, and expect to receive it.

The Gherkin is both a response and contributor to its environment – the need to maintain historic view corridors and protect the street level from downdrafts fashioned a most unlikely icon, without which the London skyline already seems unthinkable. The Gherkin still stands out as a building admired by both occupiers and community. The quality of construction and the building's environmental responsiveness mean that the Gherkin will continue to be an icon in the City of London for years to come.

> *"You can tell a skyscraper has made it into the subconscious of a city when the taxi drivers emote about it. After ten years this building still inspires us – it says a lot about the power of tall buildings to contribute to their urban environments and add to the evolving identity of a city."*
>
> Jeanne Gang, Jury Chair, Studio Gang Architects

offices to entirely open-plan floors, persists today. The widening and slimming profile generates a variety of floor plates that can respond to different sectors and markets.

The building is exemplary in terms of environmental and energy performance. The natural ventilation system operates by importing external air into the building through building management system (BMS)-controlled, motorized perimeter windows placed in each of the six light wells. The adoption of natural ventilation varies, depending on tenant layout and requirements. Approximately 50 percent of occupants currently use the system.

An active, ventilated façade is used across the whole building. This comprises a low-emissivity, double-glazed clear external unit to the outside and a single-pane interior glass, separated by a ventilated cavity. Within the cavity are solar control blinds operated by the BMS. A proportion of office extract air is passed through the façade cavity, which takes the intercepted heat reflected by the blinds from the façade back to the outside via on-floor air handling units. This minimizes solar gain in the offices and makes the façade effectively part of the office extract system.

The pitch angle of the blinds is fixed by individual, BMS-controlled dedicated motors to an optimum position to reduce solar gain within the office spaces at all times, while maximizing light transmission through the gaps in the blinds. Ten years on, this system is

operational and effective in providing user comfort, while reducing energy demand.

The Gherkin is not just an icon; it also provides a contribution to the urban realm beyond itself. The outdoor space is another great success of the project, where the building's contribution to the city has been most evident: the plaza is full of people in the summer, with food markets, city events, and a dynamic arts program illustrating its success.

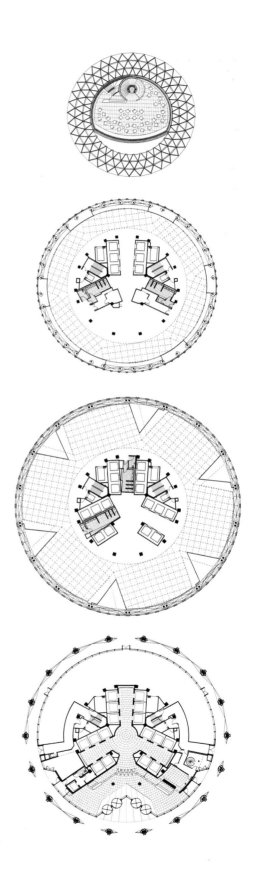

Opposite: Interior view of the apex of the building at the 40th floor

Above: Views of the active public plaza surrounding 30 St Mary Axe

Right: Comparison of tapering floor plans (from top to bottom) – 40th-floor private bar level, 33rd- and 21st-floor office levels, and ground floor

Co-Winner
CTBUH Innovation Award

BSB Prefabricated Construction Method
Used in T30 Hotel, Changsha

Innovation Design Team:
BROAD Group

Opposite: Workers assemble the prefabricated components on-site

Above: Overall view of T30, Changsha, where the BSB Prefabricated Construction Method was utilized

"You can see the mechanical engineering that permeates through this comprehensive innovation; as a group they have great potential for further development and innovation."

David Scott, Juror, Laing O'Rourke

Tall building construction has always been a time-consuming enterprise, and it amplifies the widely acknowledged contribution of the building industry to landfill-bound waste. It is also a significant challenge to limit the energy consumption of massive buildings once constructed.

In observation of this challenge, BROAD Group devised the Broad Sustainable Building (BSB) technology, a prefabricated construction method that adopts modular design technology to vastly reduce the waste associated with construction. The aim of this process is to lower energy consumption, reduce environmental pollution, and save energy resources. The ultimate benefit will be the industrialization of the construction industry. BSB makes use of fabricated, diagonally braced joint steel frame structures, which offer resistance to earthquakes of a magnitude of 9 on the Richter scale, are 90 percent factory assembled, and produce buildings that contribute only one percent of the construction waste

of a comparable conventionally built structure. From a structural perspective, the BSB system distributes stresses evenly and supports a high bearing capacity. The diagonally braced reinforced joint steel frame structure system consists of a main board, columns, and diagonal bracing.

Construction members are factory made. The floor slab is made up of a concrete-filled, profiled steel sheet, which is affixed to steel beams – this creates a "board" module. Columns support the board; diagonal bracing is set between beams and columns. Heavy construction members are joined by high-strength bolts on-site. This approach helps to achieve the purpose of easy installation, easy disassembly, and easy maintenance, while still ensuring strength.

> "This project defines why the Innovation Award was created. The project advances the thinking about what is possible in the areas of pre-fabrication and speed of assembly."
>
> Richard Cook, Juror, COOKFOX Architects

The BSB project has changed the construction mode from an extensive construction process into a lean, sustainable production line. Only ten percent of construction time is on-site; the rest is inside the controlled conditions of a factory. Unlike conventional construction sites, in the factory there is virtually no risk of fire, no water or dust infiltration, no noxious odors, and no construction waste. Not only is environmental pollution reduced; the quality of construction, increase in productivity, and lowered cost of construction has significant implications for the future of building tall. On-site, workers lift boards, tighten bolts, and conduct painting. No sawing or welding is required, reducing the time spent, and noise

Opposite Left: To-scale seismic test model of the modular structural system at BROAD Group's testing facilities in Changsha, China

Opposite Right: Recently completed modular components are stored in a warehouse before delivery on-site

Right: Interior view of one of the modular structural components

Jury Statement

Tall building designers have used prefabrication techniques on discrete elements for years, but never before has an entire prefabricated building system been developed to this degree. It is both a structural and mechanical engineering response to the demands of a rapidly urbanizing world. Integrating a bolted assembly technique with triple glazing, automatic blinds and air filtration systems, the BSB Method is a clear and innovative way of fundamentally rethinking tall building construction, and holds great promise for the future.

and traffic disruption to neighboring communities. Beyond the improvements granted to basic structural erection, the BSB system employs more than 30 energy-saving technologies in interior and exterior assemblies. While three centimeters of thermal insulation is standard in China, BSB uses 20 centimeter panels and four-paned windows.

The mechanical systems are also pre-assembled at BROAD Group's facilities. The heat-recovery fresh air system exchanges heat between outdoor fresh air and indoor exhaust air, recovering between 70 and 90 percent of the expended energy, providing extremely high indoor air quality.

The air filtration system is also prebuilt in the factory and optimized for areas with poor air quality, such as China's urban regions. The initial stage of the system uses typical coarse filters that collect the largest particles. The second stage uses the Broad-specific electrostatic cleaner. Remaining particulates are removed in stage three by a HEPA filter. The final filtration efficiency can be as high as 99 percent. In a residential application, each room is equipped with a monitor that can check particulates, formaldehyde, and CO_2 levels at any time, while comparing with outdoor conditions.

The process has been successfully implemented through the construction of the T30 Hotel, a 30-story, prefabricated hotel building in Xiangyin, Hunan Province. This construction project caught the world's

attention through the distribution of a YouTube video depicting its construction in just 15 days.

BROAD Group has even higher ambitions for the technology: it has proposed to build the world's next tallest building (at 838 meters) using the BSB Method in just nine months.

Co-Winner
CTBUH Innovation Award
KONE UltraRope™

Innovation Design Team:
KONE Corporation, Ltd.

Opposite: KONE UltraRope in a high-rise elevator shaft
Above: KONE UltraRope in the elevator hoisting machine

"It is not an exaggeration to say that this is revolutionary, but it is not just the enablement of greater height that is beneficial – the greater energy and material efficiencies are of equal value."

Antony Wood, Juror, CTBUH

The KONE UltraRope™ is a new hoisting technology that could eliminate many of the disadvantages of conventional steel rope, and opens up new possibilities in high-rise building design – an important consideration as urbanization brings increasing numbers of people to cities.

Comprised of a carbon-fiber core and a unique high-friction coating, KONE UltraRope is extremely light, which reduces energy consumption in high-rise buildings. The drop in rope weight means a dramatic reduction in elevator moving masses – the weight of everything that moves when an elevator travels up or down, including the hoisting ropes. Due to the significant impact of ropes on the overall weight of elevator moving masses, the benefits of KONE UltraRope increase as travel distance grows.

Elevators are currently limited to a single-shaft height of approximately 500 meters, requiring transfers to reach the top of supertall buildings. Because steel rope's

Finalist
CTBUH Innovation Award

Megatruss Seismic Isolation Structure
Used in Nakanoshima Festival Tower, Osaka

Innovation Design Team:
Architect/Structural Engineer: Nikken Sekkei, Ltd.

Opposite: Megatruss members visible in the 13th-floor sky lobby
Above: Overall view of the Nakanoshima Festival Tower, Osaka, where the megatruss seismic isolation structure was utilized

"Seismic isolation of tall buildings was never an option, until now. This system allows towers to be built over large open spaces, in areas of severe seismicity with all the benefits of seismic isolation."

David Scott, Juror, Laing O'Rourke

The seismically isolated structural system is one of the most effective technologies used in countries frequently hit by earthquakes. The prevailing notion holds that such isolation systems are not commonly adaptable to a high-rise building. High-rise buildings have long-period motions; in order to incorporate seismic isolation at the base, the tower above must be extremely slender – often too slender to be economically practicable. Mid-story seismic isolation, when combined with megatrusses, can avoid this issue.

The only working example of this solution can be found in the Nakanoshima Festival Tower in Osaka, Japan, a multi purpose high-rise building complex with about 145,600 square meters in total floor area, and a height of 200 meters. The program consists of a 2,700-seat music hall topped by offices. These sections are seismically isolated from each other in two places – at 45 meters above ground, forming the roof of the auditorium seating in the Festival Hall, and at 54 meters, at the roof of the Festival Hall stage (see section on next page).

Left: Section diagram
Opposite Top: Lead rubber bearings (LRBs) on the seismic isolation floor
Opposite Bottom: Megatruss and belt-truss concept diagram

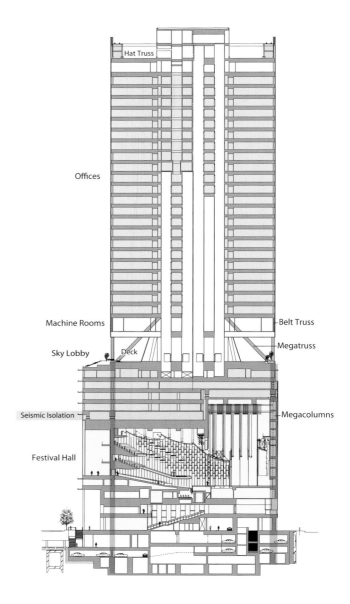

Jury Statement

The megatruss seismic isolation structure provides a way forward for some of the most densely urbanized, and urbanizing, populations, many of which happen to lie across active seismic zones. This solution affords levels of programmatic flexibility that would have been prohibitively expensive to build safely using other methods, while discretely addressing the problem of occupant comfort during swaying. This may accelerate the trend of mixed-use vertical cities along the Pacific Rim and elsewhere.

In general, concert halls are best constructed out of reinforced concrete walls to form a rigid frame that supports sound isolation and acoustic performance. By contrast, offices are ideally column-free for maximum flexibility. The intermediate seismic isolation enables these contrasting requirements to coexist in the same building.

Steel megatrusses support the center columns of the high-rise office floors and form the roof of the hall space below. These are surrounded by a belt truss, which connects megacolumns on the outer perimeter of the podium. The diagonal members of the megatruss meet the megacolumns below the belt truss. The belt trusses convey approximately 400,000 kN of force from 128 office-level columns to the 16 megacolumns surrounding the concert hall, such that each megacolumn carries approximately 60,000 kN of force.

The diagonal members of the megatruss are parallelograms in section, measuring about 1,200 by 850 millimeters. Joints of the megatrusses, each weighing a maximum of 60 metric tons, are mostly weld-assembled on-site. On the sky lobby floor, the diagonal members of the megatruss and the megacolumns are exposed to occupants' view.

Lead rubber bearings (LRBs) and oil dampers are installed on the the seismic isolation floor of this building, and two sets of square-shaped LRBs, at 1,500 millimeters long (the largest in Japan), are joined together to support the megacolumns. By virtue of the megatrusses, the majority of the force imposed by the peripheral columns is sent to the isolation devices, delivering very high performance as a seismically isolated building.

To assess the unprecedentedly large axial force acting on the columns, the designers employed two sets of seismically isolating devices connected to each other for each column base, and also employed the square-shaped laminated rubber bearing, which has a larger bearing capacity. For the connection of megatruss members, which required complicated detailing, the shape was determined by three-dimensional analyses and fabrication studies.

The intensive studies resulted in companion solutions. For instance, elevators that pass through seismically isolated floors have flexible rails to accept the action of the seismic-isolation mechanism. Seismically isolated floors are equipped with stoppers to prevent them from swaying during wind storms.

> *"This solution brings an innovate approach that is made clearly visible to people through the powerful diagonal columns of the megatruss cutting through the inside of the sky lobby."*
>
> Karen Weigert, Juror, Chicago Chief Sustainability Officer

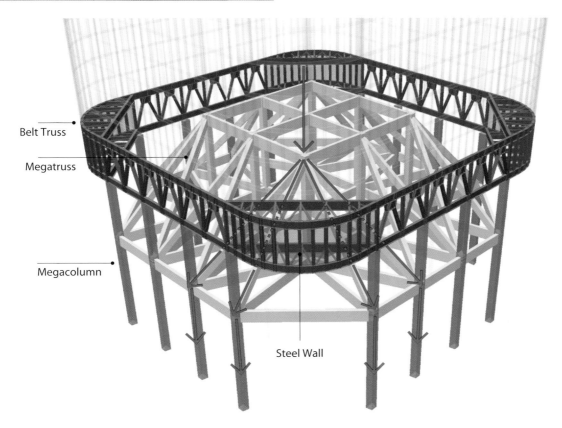

- Belt Truss
- Megatruss
- Megacolumn
- Steel Wall

Finalist
CTBUH Innovation Award

Precast Concrete Façade
Used in Tour Total, Berlin

Innovation Design Team:
Architect: Barkow Leibinger
Structural Engineer: GuD Planungsgesellschaft für Ingenieurbau mbH
Fabricator and Façade Construction (concrete): Dreßler Bau GmbH
Façade Construction (metal): FKN Fassadentechnik GmbH & Co. KG

Opposite: Close-up view of the precast concrete façade
Above: Façade panels awaiting on-site delivery

"The creative integration of the precast load-bearing façade is a good example of how prefabricated solutions can support an elegant and effective building design."

Robert Okpala, Juror, Buro Happold

The "raster" façade used on the Tour Total building in Berlin is a load-bearing precast concrete frame that eliminates interior columns, allowing floor-to-ceiling glass by way of triple-glazing with exterior retractable protective louvers, and generates more usable floor area than other systems. The precast façade generates a ratio of 60 percent glazed to 40 percent closed surfaces, improving insulation values. Precast concrete is also an inherently fireproof building material, eliminating the need for additional fire protection. This also means that the building requires only a single exit stair with baffle and sprinklers, generating additional usable floor area.

The basic module is a two-story, T-shaped precast element. These elements stagger and overlap from floor to floor. In this application at Tour Total, Berlin (see Best Tall Building Europe section where it was awarded Finalist status, page 126), each precast element is then "creased" with a K-shaped, origami-like fold, using the full depth (up to 50 centimeters) for each piece. When combined, the overall effect produces a visual diagonal

continuity, which wraps around the corners of the building. This is an effect that emphasizes the vertical, exaggerating the height of a high-rise building. This deep façade protects the interior from direct sunlight, and also provides a pocket depth to integrate retractable louvers for additional sun control.

This type of construction will remain relevant for future construction for a number of reasons. Digital fabrication technologies will continue to evolve, and will have a direct impact on how formwork is made for concrete. Historically, precast has worked best with a high degree of repetition or modulation of individual elements. From a design point of view, in the future this need not be the case. Digitally fabricated formwork, or formwork that is adaptable from pour to pour, soon will be a reality and will allow a great deal of variation in the number of individual building elements. Of course, this influences aesthetic/formal articulation, but also, to a larger degree, allows the structure to respond more precisely (and economically) to structural and loading variations, so that a building can reflect those conditions and change from floor to floor. In Tour Total there are some 1,395 precast façade elements, using approximately 41 individual cast types. These innovations are in the context of a developer building with constraints in both budget and allowable construction time. At 840 Euros per square meter, this is a very economical façade and construction technique for Berlin and beyond.

Prefabricated concrete is well positioned to replace steel for functional, sustainable, aesthetic, and economic reasons. It is also a faster way to build, as precast concrete elements require no additional handling for fireproofing or finishes. When the elements arrive on-site, they need only be assembled.

With a tolerance of 2-3 millimeters, precast concrete is a remarkably precise material that needs no further finishing and is incredibly consistent in terms of color and finish. The precision is significant, but aesthetically the plasticity of concrete will continue to improve,

Jury Statement

The next evolution of precast concrete façade systems provides efficiencies in cost, construction, and environmental performance. Reminiscent of the precursor to skyscraper wall technologies – the load-bearing masonry wall with punched openings – the raster façade retains its environmental benefits, while taking a leap into the future in terms of ease of fabrication/installation. The raster façade system ameliorates the wasteful cladding practices of the 20th century while launching a time-tested approach into the 21st.

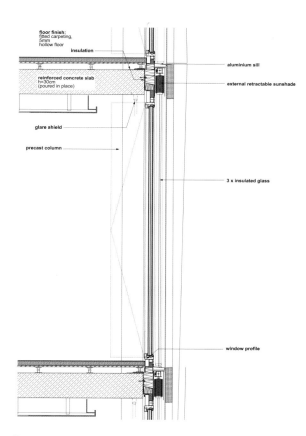

Opposite: Casts for prefabrication of façade units

Left: Detail façade section

Below: Diagram of prefabricated structural panels, insulation/glazing, then prefabricated decorative panels (top), and construction view showing installation of decorative panels onto the structure (bottom)

> "The fully precast load-bearing façade is rich in texture and detail and is used to create an energy-efficient rippling skin, that is dramatic and exciting when viewed at a distance or close-up."
>
> David Scott, Juror, Laing O'Rourke

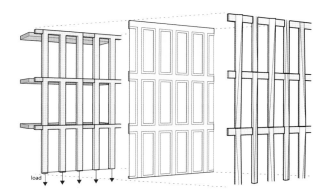

offering façades with sculptural depth that control light, radiation, and weather protection. This offers an intelligent alternative to the energy-wasting and impractical current use of the all-glass curtain wall.

Raster façades will become more practical with improvements to the thermal connections back to the floor slabs. Tour Total is a sandwich construction that consists of an outside layer of precast concrete, which is both ornamental and stiffens the grid against wind load, a layer of thermal insulation and glazing, and an inner structural grid of precast poured as "combs." All three layers are fastened together on-site. In the future, the inner layer will be eliminated, as "iso-corb" connections are improved, allowing for an insulated structural connection between the "cold" precast and the "warm" interior structural floor slabs. These connections, as they improve, can allow for thermal expansion in tall buildings. The advantage of this emerging technology will be to increase usable floor area and speed up the construction process and time.

Finalist
CTBUH Innovation Award

Rocker Façade Support System
Used in Poly Corporation Headquarters, Beijing

Innovation Design Team:
Architect/Structural Engineer: Skidmore, Owings & Merrill LLP
Associate Architect: Beijing Special Engineering Design and Research Institute

Opposite: The Rockers in foreground support the cable-net glass wall behind
Above: Overall view of the Poly Corporation Headquarters, Beijing, where the Rocker system was utilized

"As the largest single-layer cable-net façade system in the world, it has helped the building achieve an integration of Chinese traditional culture and modern high-tech application."

Nengjun Luo, Juror, CITIC Heye Investment

The Poly Corporation Headquarters features a unique design component that was created specifically for this building – The Rocker, which supports the world's largest cable-net glass wall while actively releasing the effects of earthquakes and heavy winds. Additionally, it facilitates the suspension of an eight-story, lantern-like museum structure within the office building's atrium.

Structural analysis showed that the support for the 22-story-tall glass atrium wall could not be reasonably achieved using a conventional two-way cable net, but could be achieved if the 90-meter-high by 60-meter-wide enclosure was broken down into smaller segments. A cable-stayed system was introduced by using two large-diameter parallel strand bridge cables in diagonal fold lines while anchoring to the eight-story suspended museum structure. The museum structure acts as a counterweight for the cables, introducing pre-stress and providing the required stiffness to resist out-of-plane loads caused by wind on the cable net. In addition to the diagonal cables used at the atrium glass wall, two

> *"There is a lightness and delicacy about this tensile façade that belies the huge engineering achievements – which are also prominently on display."*
>
> Antony Wood, Juror, CTBUH

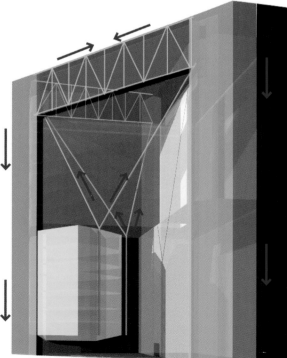

additional cables and a Rocker were introduced at the rear of the museum structure to assist in its suspension.

Extensive study showed that when installed straight, the diagonal cables curved toward the façade during tensioning of the cable net, reacting against the faceted form of the net. The length of each connecting rod between the diagonal cables and the net was pre-calculated to result in planar glass surfaces at the end of all stressing operations. Using the faceted design solution, the typical horizontal and vertical cables are minimized at 34 and 26 millimeters, respectively.

By diagonally connecting the structure at the roof of the lantern to the structure 10 stories higher at the top of the atrium, bridge cable elements would act as mega-braces when the structure was subjected to earthquake loads. Up to 900 millimeters of relative movement between the two roofs is expected in an extreme seismic event. If conventionally connected to the base building structure, earthquake-induced forces would have resulted in the cables and their connections becoming unfeasibly large.

Recognizing that the building sways sideways in multiple directions during an earthquake, the team developed an innovative mechanism that would manage these demands, while introducing no additional force into the base building structure. When the cable-net façade sways, one of the two diagonal cables lengthens as the other shortens, with the rocker acting as a "reverse pulley," using steel castings/plates designed to pass

between one another and interconnect with steel pins allowing for movements in both building directions. Because the bridge cable sizes were so large (over 200 millimeters in diameter) a conventional drum-type pulley would have required a diameter of 6 meters.

To test the Rocker, a working model was created to simulate the expected building movement. Springs were connected at the ends of the cables. If they did not elongate during the side-to-side motion, additional forces within the cables would not be present. The experiment found that no significant elongation occurred during movement, confirming that the concept would work and could be installed in the first application of its kind. This innovation could be adapted for use in other situations that might be encountered in unique high-rise buildings, where brace-like members might need to be freed from participation in the lateral systems of the structures.

Opposite Top: Close-up view of a Rocker
Opposite Bottom: Structural loading diagram
Above: Looking up at the museum structure within the cable-net glass wall

Jury Statement

The best structural solutions concentrate intense study and experimentation into an application that simultaneously looks effortless and near-impossible. The Rocker Façade Support System floats an expansive glass wall astride a massive suspended escarpment, in a balancing act so awe-inspiring that it suggests optical trickery. It forges the way for tall buildings to include faceted glass walls that maintain openness in the interior, but are structurally sound enough to withstand extreme seismic events.

CTBUH Lifetime Achievement
Awards Criteria

Lynn S. Beedle Award
The award recognizes an individual who has made extraordinary contributions to the advancement of tall buildings and the urban environment during his or her professional career. These contributions and leadership are recognized by the professional community and have significant effects, which extend beyond the professional community, to enhance cities and the lives of their inhabitants. The individual's contributions may be well known or little known by the public and may take any form, such as completed projects, research, technology, methods, ideas, or industry leadership.

The candidate may be from any area of specialization, including, but not limited to: architecture, structure, building systems, construction, academia, planning, development, or management. The award emphasizes the unique, multi disciplinary nature of the Council and is thus set apart from other professional organizations' awards for single disciplines.

Fazlur R. Khan Medal
The award recognizes an individual for his/her demonstrated excellence in technical design and/or research that has made a significant contribution to a discipline(s) for the design of tall buildings and the built urban environment. The contribution may be demonstrated in the form of specific technical advances, innovations, design breakthroughs, building systems integration, or innovative engineering systems that resulted in a practical design solution and completion of a project(s). The consideration may be based on a single project or creative achievement through multiple projects.

In the case of both Lifetime Achievement Awards, the candidate may or may not be a member of the Council. The contributions of the award recipients should be generally consistent with the values and mission of the CTBUH and its founder, Dr. Lynn S. Beedle. The awards are not intended to be awarded posthumously, although they may be so awarded in some cases. The two Lifetime Achievement Awards are selected by the CTBUH Board of Trustees.

Winner
Lynn S. Beedle Lifetime Achievement Award

Henry N. Cobb
Pei Cobb Freed & Partners

Opposite: Hancock Place, Boston, 1976 (241 m / 790 ft)
Above: Henry N. Cobb

"Henry Cobb has demonstrated phenomenal leadership throughout his career, particularly on Boston's Hancock Place, which was a hugely important moment for the tall building industry."

Timothy Johnson, CTBUH Chairman, NBBJ

The defining characteristic of Henry N. ("Harry") Cobb's career has been his passionate reconsideration of the tall office building as a presence in the city, which Cobb calls "Skyscraper as Citizen." From One Place Ville-Marie in Montreal (completed in 1962) to the Palazzo Lombardia in Milan (completed in 2011), Cobb's numerous office towers consistently exhibit his ongoing preoccupation with the question of how tall buildings can shape rather than merely preempt the space of the city.

In keeping with his belief that architecture is above all an art of place making, he has envisioned the office tower not as an autonomous object, but rather as a contingent presence responsive to the uniqueness of its specific place in the city.

"For me, the way a tall building meets the ground has always been at least as important as the way it meets the sky," Cobb has said.

Left: Place Ville-Marie, Montreal, 1962 (188 m / 617 ft). One of Cobb's early high-rise projects, it was distinctive for the positive contribution made to the urban ground plane.

Opposite: Fountain Place, Dallas, 1986 (219 m / 720 ft). A water garden at the base of the tower ties this prismatic form to the urban environment.

Born and raised in Boston, he was educated at Phillips Exeter Academy, Harvard College, and Harvard Graduate School of Design, from which he received his Master of Architecture degree in 1949. After a brief stint working for Hugh Stubbins in Boston, he moved to New York to join I. M. Pei in his fledgling practice under the auspices of the legendary developer William Zeckendorf, Sr. In 1955, together with their colleague Eason H. Leonard, Pei and Cobb founded I. M. Pei & Associates, now Pei Cobb Freed & Partners.

The 1962 One Place Ville-Marie in Montreal (formerly the Royal Bank of Canada Building) – an example of Henry's early high-rise work – displays many of the characteristics of other International Style buildings of the early 1960s, but it is distinctive in the way that it knits together urban life in the air, at the surface, and underground. Nearly half of its area is below grade, forming the nexus of Montreal's underground walkway system, protecting citizens from the harsh winters while plugging the vibrancy of center city life into the building's interior, with what architectural historian Mark Pimlott calls "episodes of civic gravity and monumentality."

Hancock Place in Boston cemented Cobb's reputation as a sophisticated architect, though the problematic project could just as easily have undone the career of a less ethical designer. The headquarters of the John Hancock Mutual Life Insurance Company employed an unusual rhomboid shape, covered entirely in reflective glass, so as to fit into an awkward site and

Trustee Statement

Henry Cobb's career is remarkable because of his ethical and forward-thinking approach to the design of tall buildings and their place in the urban fabric. Transforming the efficiencies of International Style into contextually responsive projects that nevertheless define the skyline of the cities in which they were built, Cobb not only raised the standards of design for tall buildings, but of the building profession in general. His professional composure and deep knowledge have set the template for future tall designers.

reduce impact on neighboring historic structures, most notably Trinity Church. During an arduous eight-year construction period from 1968 to 1976, the building endured a series of mishaps, the most notorious of which was the failure of insulating glass units, which necessitated the removal and replacement of all 10,334 panels in the curtain wall. In an act of candor almost unimaginable in today's liability-tinged building environment, Cobb personally informed the Boston Building Commissioner of his finding that the insulating glass panels were defective, and that he had directed that they be removed.

"Harry showed phenomenal leadership on that project," said William Baker, CTBUH Trustee and structural engineering partner at SOM. William LeMessurier, a structural engineer who worked on the Hancock Tower project with Cobb, told Architecture magazine in 1988, "Harry Cobb's performance was not only responsible, it was inspiring. . . .Whenever I have had some problems in my own professional life that made me have to stand up and be responsible for my client's interest, I said I will have to behave like Harry."

Cobb has continued to design projects that exemplify technological advancements and best contemporary practices through the decades, but always found ways to connect smoothly sculptural, shining icons to the daily life of cities at the ground plane. Fountain Place in Dallas not only achieves a distinctive prismatic identity on the skyline; it also has a water garden flowing through its base. Tour EDF at La Défense,

Opposite Left: Tour EDF, La Défense, Paris, 2001 (148 m / 486 ft). A large disk-shaped canopy marks the entrance to this tower.

Opposite Right: Torre Espacio, Madrid, 2008 (224 m / 735 ft). The tower's seemingly twisting form comes down to create a protected plaza.

Right: Palazzo Lombardia, Milan, 2011 (161 m / 529 ft). Recipient of the CTBUH 2012 Best Tall Building Europe award, the slender tower houses government offices, and the low-rise "strand" buildings weave across the site to create pockets of public space and an enclosed public plaza.

Paris, splits its prow-like profile to draw the eye to a generous canopy over the pedestrian entrance. Torre Espacio in Madrid appears to twist through an organically nonlinear turn to shelter its surrounding plaza. Palazzo Lombardia in Milan represents an important investment in the civic realm, by providing not only a sleek, light-filled, narrow tower for the local government offices, but also a linear public park and glass-enclosed central plaza, tying together an auditorium, exhibition space, and restaurants.

Beyond designing tall buildings, Cobb has coupled his professional activity with teaching. He has lectured widely and has held visiting appointments at a number of universities. From 1980 to 1985, he served as Studio Professor and Chairman of the Department of Architecture at the Harvard Graduate School of Design, where he continues to teach occasionally as a visiting lecturer. In 1992, he was Architect in Residence at the American Academy in Rome.

Cobb is a Fellow of the American Institute of Architects, a Member (currently President) of the American Academy of Arts and Letters, a Fellow of the American Academy of Arts and Sciences, and an Academician of the National Academy of Design. Awards recognizing his achievements as both architect and educator include the Arnold W. Brunner Memorial Prize in Architecture and the Topaz Medallion for Excellence in Architectural Education. He has received honorary doctorates from Bowdoin College and the Swiss Federal Institute of Technology.

Winner
Fazlur R. Khan Lifetime Achievement Medal

Clyde N. Baker, Jr.
AECOM

Opposite: Willis (formerly Sears) Tower, Chicago, 1974 (442 m / 1,451 ft). Clyde Baker has been involved in the geotechnical engineering for six of the 12 tallest buildings in the world, including Willis Tower.

Above: Clyde N. Baker, Jr.

"Without Clyde Baker's work, the building on the top doesn't mean anything. Every time I encounter a big problem with a high-rise building foundation, the first person I want to call is Clyde."

Dennis Poon, CTBUH Trustee, Thornton Tomasetti

Clyde N. Baker, Jr. has performed geotechnical engineering for six of the 12 tallest buildings in the world, and a major portion of the high-rise buildings built in downtown Chicago over the past 50 years, including nine with deep basements and slurry walls.

In Chicago, he has served as geotechnical engineer or consultant on the Willis (formerly Sears) Tower, the John Hancock Center, Trump Tower, the Aon Center, Water Tower Place, 900 North Michigan, and the AT&T Center. Outside the United States, he has performed engineering work on the National Commercial Bank, Jeddah, Saudi Arabia; Petronas Towers, Kuala Lumpur, Malaysia; and Taipei 101, Taiwan. Mr. Baker was a peer reviewer on the Burj Khalifa, Dubai, UAE; the T & C Tower in Kaohsiung, Taiwan; and the Chicago Spire, a 2,000 foot (600 meter) tower. The Chicago Spire was planned for Chicago's lakefront before construction stopped due to the economic recession of 2008. Had it been built, it would have held the record for the highest soil-bearing pressure foundation in the world upon its completion.

Left: John Hancock Center, Chicago, 1969 (344 m / 1,128 ft)

Opposite Top: 900 North Michigan Avenue, Chicago, 1989 (265 m / 869 ft)

Opposite Bottom: 200 Public Square (formerly SOHIO Building), Cleveland, 1985 (201 m / 658 ft). Baker developed a unique friction caisson foundation design for this structure.

At one of his earlier works, 200 Public Square, Cleveland, USA, formerly known as the SOHIO Building, Mr. Baker undertook two substantial tasks. He developed a unique friction caisson foundation design for the 46-story structure, which were the deepest known caissons in the USA to date (the building was completed in 1985). Work involved a full-scale, instrumented caisson load test to confirm friction design parameters and included instrumentation and long-term monitoring of a main production caisson. In addition, on this project Mr. Baker developed a non-destructive testing evaluation program, to permit caisson construction and concrete placement entirely under water.

At the Petronas Towers, Mr. Baker developed and provided oversight for a successful limestone cavity filling and slump-zone grouting program below the world's deepest building foundations, which extended below the water table.

He also developed a controlled displacement plan for the organic marine silt at the large Sha Tin Land Reclamation project in Hong Kong. Marine silt was displaced into local silt ponds, leaving silt-free corridors for roadway and utility support, with greatly reduced settlement problems. High-rise structures were constructed on pile-supported foundations in the filled-over silt ponds, eventually forming the Sha Tin New Town.

More recent projects that Clyde has been involved in which are currently under construction include the

Trustee Statement

Through a career spanning 59 years, Clyde Baker has become synonymous with geotechnical engineering for tall buildings, devoting passionate energy and innovation to the practice. In so doing, he has not only expanded the scope of understanding around a complex engineering subject; he has provided a model of professional ethics from which others can draw. By "shirking" the professional practice "learnings" that we are all getting, which amount to saying, "that's not my fault," Clyde Baker put the profession at the heart of his interest, not the lawyers.

world's next tallest building, Kingdom Tower, Jeddah; Lotte World Tower, Seoul; Busan Lotte Town Tower, Busan; and the Doha Convention Center and Tower, Doha.

As a result of his experience, Mr. Baker has developed an international reputation in the design and construction of deep foundations. He has been a leader in using in situ testing techniques, correlated with past building performance, to develop more efficient foundation designs. He has also been instrumental in increasing the use of the Menard Pressuremeter, which monitors the relationship between pressure and deformation in soil, in the United States. In Chicago's soil, Mr. Baker has mastered the economical use of belled caissons on hardpan soil for major structures in the 60- to 70-story range, which normally would have required extending the caissons to solid rock at a significant cost premium.

In an industry increasingly constrained by fear of litigation, Mr. Baker has also shown extraordinary leadership in the field, making project-critical decisions "on the spot" based on his own site observations and experience. CTBUH Trustee William F. Baker (no relation), structural engineering partner at Skidmore, Owings & Merrill, recounted a scenario in which the Chicago Tunnel and Reservoir Plan (TARP) project had pushed the water table up, causing unexpected flooding at a downtown tall-building project site under Clyde Baker's purview. Judging the situation quickly, Mr. Baker determined that the flooding would not be catastrophic

Left: Taipei 101, Taipei, 2004 (508 m / 1,667 ft)

Opposite : Petronas Towers, Kuala Lumpur, 1998 (452 m / 1,483 ft). Baker developed and provided oversight for a successful limestone cavity filling and slump-zone grouting program below the world's deepest building foundations.

if the caissons were filled with concrete faster than they could fill with water.

"He began directing concrete trucks himself," Bill Baker recounted. "He did exactly what the insurance companies tell you not to do. But if he had sat back and kept quiet, then there would have been a problem."

Clyde Baker received his BS in Physics from William & Mary College in 1952, and his BS and MS degrees in Civil Engineering from Massachusetts Institute of Technology in 1952 and 1954, respectively. He joined the staff of STS Consultants, Ltd. (formerly Soil Testing Services) in the fall of 1954, eventually rising to the role of chairman. STS was acquired by AECOM in 2007, and Mr. Baker served as technical director of AECOM's geotechnical practice until his official retirement in 2013, though he continues to consult with them.

Mr. Baker has shared his knowledge and experience with his peers through numerous Conference and University lectures, technical articles, papers, and publications. He is the recipient of the Deep Foundations Institute Distinguished Service Award, the ADSC Outstanding Service Award, ASCE's Ralph B. Peck, Thomas A. Middlebrooks, and Martin S. Kapp awards. He has received the 2007 Award of Excellence from Engineering News-Record magazine. He also received three Meritorious Publication Awards from the Structural Engineers Association of Ilinois (SEAOI), including "The History of Chicago Building Foundations, 1948 to 1998" and is the author of "The Drilled Shaft Inspector's Manual" sponsored jointly by the Deep Foundation Institute and the International Association of Foundation Drilling (ADSC).

Mr. Baker is an Honorary Member of ASCE, a past President of SEAOI and the Chicago Chapter of ISPE, a past Chairman of the Geotechnical Engineering Division of ASCE, a past Editor of the Geotechnical Engineering Journal, and a past Chairman of ACI Committee 336 on Footings, Mats, and Drilled Piers. He is a member of the National Academy of Engineering and in 2006 received The Moles 2006 Non-Member Award for "Outstanding Achievement in Construction." He was also the winner of the most prestigious award in geotechnical engineering, giving the Terzaghi Lecture in 2009.

CTBUH 2013 Fellows

CTBUH Fellows are recognized for their contribution to the Council over an extended period of time, and in recognition of their work and the sharing of their knowledge in the design and construction of tall buildings and the urban habitat.

Ahmad Abdelrazaq
Samsung C&T Corporation, South Korea

Ahmad Abdelrazaq is Senior Executive Vice President and the head of the High-rise and Complex Building Division at Samsung C&T Corporation, Seoul, Korea. Since joining Samsung in 2004, Mr. Abdelrazaq has overseen the Division transition from a traditional construction-only provider into the successful design-build, pre-construction, value engineering, and fast track design/construction for high-rise and complex buildings.

Mr. Abdelrazaq has been involved in CTBUH activities worldwide over a number of years. He currently serves with the CTBUH as an Advisory Group Member, CTBUH Journal Editorial Board Member, and has spoken at many CTBUH Conferences. He has previously been a member of the CTBUH Awards Jury, and contributed to the CTBUH Technical Guide on Outrigger Design for High-rise Buildings.

"Ahmad is a noted figure and has been very supportive of CTBUH through the years," said CTBUH Chairman Timothy Johnson. "He is an integral presence."

Felino A. Palafox Jr.
Palafox Associates, Phillippines

Felino Palafox, Jr., is widely appreciated as one of CTBUH's most energetic members. As the Philippines Country Representative since 2009, Palafox has attended almost all recent CTBUH Conferences. Palafox founded Palafox Associates, an architectural design and planning firm in the Philippines, which for the past 22 years, he has led and managed.

Palafox is the first and only architect, urban planner, and environmental planner to be elected President of the Management Association of the Philippines, a 60-year-old management organization whose 740 members represent a cross section of CEOs, COOs, and other top management practitioners from the largest local and multinational companies in the Philippines.

"[Palafox] presents a different kind of energy than just being on a panel or committee," Johnson said. "He represents a good benchmark and way to think about what this particular title is about."

CTBUH 2013 Awards Jury

Jeanne Gang, 2013 Awards Jury Chair | Studio Gang Architects | Chicago, USA

Jeanne is Founder and Principal of Studio Gang Architects, whose inventive tall buildings respond locally (site, culture, people), as well as resound globally (density, climate, sustainability). Her Aqua Tower in Chicago brings sustainability and a sense of community to urban living and was awarded as a CTBUH Tall Building Awards Finalist in 2010.

Richard Cook, COOKFOX Architects | New York, USA

Rick is Co-Founder and Principal of COOKFOX Architects, his work is devoted to creating beautiful, high-performance buildings. For over 25 years he has built a reputation for innovative, award-winning design. His design for the Bank of America Tower at One Bryant Park was awarded the CTBUH Best Tall Building Americas award in 2010.

Nengjun Luo, CITIC HEYE Investment Co., LTD | Beijing, China

Dr. Nengjun serves as General Manager at CITIC HEYE Investment and has been promoting the development and innovation of construction management for high-rises in China, and moving forward the use of BIM on supertall building projects from the perspective of project management during the period of construction and operation.

Robert Okpala, Buro Happold | Dubai, UAE

Robert is based in the Middle East where he is Building Environments Director for Buro Happold. He is responsible for the leadership, development, and integration of design services for the MEP and Sustainability disciplines. He is an advocate of resource efficiency and the appropriate use of low carbon, renewable technologies in the region.

David Scott, Laing O'Rourke | London, UK

David is the Lead Structural Director at Laing O'Rourke, and is past Chairman of the CTBUH (2006–2009). David has a passion for unusual structures and has extensive tall building experience that started with Foster's landmark Hong Kong Bank HQ Building. He is an advocate for low-energy buildings and off-site construction.

Karen Weigert, City of Chicago | Chicago, USA

Karen serves as Chief Sustainability Officer for the City of Chicago where she works to guide the City's sustainability strategy, bringing innovative, practical solutions to the City. She is also a producer/writer for the documentary film *Carbon Nation* which is focused on solutions to climate change.

Antony Wood, Council on Tall Buildings and Urban Habitat & Illinois Institute of Technology | Chicago, USA

Antony is Executive Director of the CTBUH, responsible for the day-to-day running of the Council. His field of specialism is the design, and in particular the sustainable design, of tall buildings. He is also a Studio Associate Professor in the IIT College of Architecture where he convenes various tall building design studios.

Review of Last Year's CTBUH 2012 Awards

October 18, 2012 marked the 11th Annual CTBUH Awards Symposium, Ceremony & Dinner, held in the Illinois Institute of Technology's S.R. Crown Hall, designed by Mies van der Rohe.

Prior to the ceremony, a day-long symposium put the spotlight on the designers, engineers, and builders involved with the recognized buildings. But the statements were not necessarily about pyrotechnics and iconography. Instead, the discussions focused more on buildings that connect to their environment and provide livable, comfortable additions to the urban fabric. "This year we were more specifically focused on the unique challenges of tall buildings," awards jury chair Richard Cook, partner at COOKFOX Architects, told the audience. Mr. Cook detailed the issues facing the award jury and the responsibility of reviewing other people's work, which he described as "soul searching." This year, for the first time, the symposium featured the work of award finalists, as well as winners.

Each presentation, with few exceptions, focused on the details and nuances of the honored buildings, free of the need to make a pitch or promote the company. Speakers detailed their shared enthusiasm for creating efficient projects that fit the urban landscape and improved their surroundings.

"Less ego and more eco," was the clarion call of Massimo Roj, design architect at Progetto CMR, the architects behind one of the featured finalist projects, Complesso Garibaldi.

"Architects don't have any ambition anymore," countered Helmut Jahn, 2012 Lynn S. Beedle Lifetime Achievement Award winner. "We are pushed into strait jackets…Now we are just pussy footing around."

While Charles Thornton and Richard Tomasetti, partners of Thornton Tomasetti structural engineering and joint recipients of the Fazlur R. Khan Medal, looked to the future, their presentation touched a nostalgic tone, as their careers chronicled many of the

biggest leaps and innovations in tall buildings in the last 30 years, growing from simple multistory projects in Florida to supertall buildings that define countries.

"It's been a real ride," Mr. Thornton said.

The ride continued into the awards dinner, which heralded the 2012 regional Best Tall Building winners: Doha Tower, Doha, Qatar (Middle East & Africa); Absolute Towers, Mississauga, Canada (Americas); One Bligh Street, Sydney, Australia (Asia & Australasia); and Palazzo Lombardia, Milan, Italy (Europe), as well as the inaugural Innovation Award and several personal lifetime achievement awards.

Absolute Towers received the award for Best Tall Building for the Americas region. The towers definitively stand out on the skyline of Mississauga, bringing distinction to the little-known, but fast-growing suburb of Toronto. At the Awards dinner, developer Joe Cordiano from Citizen Development explained

Opposite: Helmut Jahn, 2012 recipient of the CTBUH Lynn S. Beedle Lifetime Achivement Award, presents at the afternoon Symposium.

Top: Attendees network during the dinner at the 12th Annual Awards Ceremony & Dinner.

Bottom: Joint Lifetime Achievement winners, Richard Tomasetti (left) and Charles Thornton (middle) accept their Fazlur R. Khan Medals from CTBUH Chairman, Timothy Johnson (right).

Top: Absolute World Towers, Mississauga, Canada, 2012 Best Tall Building Americas winner.

Bottom: CTBUH Executive Director, Antony Wood (right) presents the Americas award to the Absolute World Towers owner representative Joe Cordiano, Cityzen Development (left) and architect Ma Yansong, MAD Architects (middle).

Opposite Left: CTBUH Awards Jury Chair, Richard Cook (far left) presents the Asia & Australasia award to members of the One Bligh Street team (from left to right): owner representative Bruce McDonald, DEXUS Property Group; architect Ray Brown, Architectus; and architect Christoph Ingenhoven, ingenhoven architects.

Opposite Right: One Bligh Street, Sydney, 2012 Best Tall Building Asia & Australasia winner.

that the vision for this project started in 2003, when his company purchased what he called "a field of dreams," where the towers would eventually be built. To answer the question, "Where are we going from here?", the company held an international design competition, attracting more than 90 submissions.

The distinct curving shape from Ma Yansong of MAD Architects clearly captured the client and the public's imagination. MAD's design stood out for its ability to offer a sharp contrast to the existing landscape of traditional rectangular buildings that had come to define Mississauga. "The Absolute Towers stretch the limits of paired sculptural form to create a marker on the skyline for a regional center," said Mr. Cook.

One Bligh Street in Sydney was awarded Best Tall Building in Asia & Australasia. The ovoid tower stood out to the jurors for its many sustainable features, including a double-skin curtain wall and a full-height, naturally ventilated atrium.

"It's difficult for Australians to feel connected to the rest of the world. This award should really go to the community and business community," Bruce McDonald, representing owner DEXUS, announced to the dinner attendees. "Australians are very proud, and will react positively to this award."

Christoph Ingenhoven, principal of ingenhoven architects, explained that while relationships between design architects and local architects are often strained, the partnership between his firm and architects of record, Architectus, was very special. "The real success of this project was the collaboration," he said.

Palazzo Lombardia in Milan won the Best Tall Building in Europe award. Recalling the mountains, valleys, and rivers of Lombardy, Palazzo Lombardia's distinctive tower is sat upon sinuous interweaving strands of linear government office space, which are seven to nine stories in height. This assemblage of gently curved glass-walled workplaces allows the building to integrate with its urban context while creating unique

landscaped public spaces that are open and inviting to the public.

"Everything comes from the aspirations of the client," lead architect Henry Cobb quietly explained at the podium. "They are fundamental. We respond to and are thrilled by them. We need them to make anything worthwhile."

On behalf of the client, the Regione Lombardia, Davide Pacca said, "The award is a symbol that gives Lombardy hope and the strength to go ahead, and gives us other reasons to go on with our projects. The aim of the project was to establish a more efficient and cost-saving space for our public institution, but also to create a place, for the city, and for the region, to host exhibitions for artists and where citizens could meet the administration and each other."

As well as the Best Tall Building Middle East & Africa award, the distinctive Doha Tower was named the overall Best Tall Building Worldwide, but only after a heated debate among jury members. Ultimately, however, the cylindrical tower in Doha, recognized for its unique shading system, won out as a "project that only makes sense in this particular place." The building was honored for incorporating elements of traditional regional design with modern technologies to create an environmentally sensitive icon for Qatar's capital.

"Doha Tower stands out for its deft and subtle sensitivity to culture, context, and climate," Richard

Cook said. "The design hints at postmodernism, but avoids this pitfall through the interpretive reuse of indigenous elements such as the mashrabiya, which varies in its density across the façade in response to solar orientation."

The overall award was announced and presented by Dutch architect Wiel Arets, the new Dean of the IIT College of Architecture. Accepting the award for Doha Tower, Hassan Al Duhaimi of developer HBS Company said, "I am very pleased to see all of you, to come to the United States," he said, proudly holding up the award. "I am thankful for everyone behind the tower."

Peter Terrell of Terrell Group also explained that the building was only made possible because the entire development team maintained its vision over pitfalls and through challenges. Their success illustrates a remarkable tribute to the client–architect relationship, ultimately creating that exceptional project.

The Al Bahar Towers in Abu Dhabi won the CTBUH's first Innovation Award for the project's computer-controlled opening-closing sun screen. The Al Bahar Towers rely on both nature and culture in the execution of their advanced screening system, which was designed to integrate the building with its cultural context and respond directly to the climatic requirements of the region. The mashrabiya form of the screens directly anchors the buildings in the Islamic tradition of the Middle East, while the dynamic movement of each of the individual units recalls the response of native plants. "I think we all know that innovation in our industry can be very difficult to achieve," Aedas Deputy Chairman Peter Oborn said upon accepting the award. "And the more encouragement we have from organizations like CTBUH, the better."

He particularly wished to point out that Abdul Majeed Karanu, a Lebanese architect, originated the concept for the dynamic mashrabiya façade. "It seems entirely appropriate that it should be someone from the region that actually came up with the specific idea," Mr. Oborn said.

The CTBUH Board of Trustees awarded the Lynn S. Beedle Lifetime Achievement Award to Helmut Jahn, the architect known for his simple and elegant designs. Never predictable, never simplistic, Mr. Jahn has developed a complex portfolio of iconic buildings around the world. Standout projects include the Sony Center in Berlin, Xerox Center in Chicago, Liberty Place in Philadelphia, and the MGM Veer Towers in Las Vegas. Once defined as a modernist, he broke away from rigid labels to create his own blends of efficient structures, paving the way for a new era of sustainable buildings.

"Helmut Jahn has created an impressive body of work that epitomizes the integration of structure, sustainability, and architecture. He is certainly deserving of the Lynn S. Beedle Lifetime Achievement Award," said William Baker, CTBUH Trustee and structural engineer at Skidmore, Owings & Merrill.

"This is not an achievement to sit on; it's an achievement to build on," Mr. Jahn told the audience.

The Fazlur R. Khan Lifetime Achievement Medal went jointly to Charles Thornton and Richard Tomasetti, founders of Thornton Tomasetti, the structural engineering firm that has helped design many of the most innovative and advanced tall buildings around the world. Their legacy includes Taiwan's Taipei 101, the Petronas Towers in Malaysia, the World Financial Center in New York City, Plaza 66 in Shanghai, and Pittsburgh's BNY Mellon Center. Equally as important, they have tirelessly given back to the community as educators and lecturers, helping to train the next generation of structural engineers.

"Charlie and Richard are one of the industry's enduring great partnerships in structural engineering, reminding us that no building is about one single person," said Timothy Johnson, CTBUH Chairman and principal at NBBJ. "By presenting this award to two people for the first time, the Council acknowledges the importance of the collaboration between two leaders who have helped build some of the world's most important structures."

"The integration of architecture and engineering, that's where the future really lies," said Mr. Tomasetti during his presentation in the afternoon symposium.

The Awards dinner also introduced three new CTBUH Fellows – Nicolas Billotti of Turner International, Georges Binder of Buildings and Data; and James Robinson of Hong Kong Land – who were honored for their ongoing contributions to the CTBUH.

Opposite Top: Palazzo Lombardia, Milan, 2012 Best Tall Building Europe winner.

Opposite Bottom: CTBUH Executive Director, Antony Wood (left) presents the Europe award to the Palazzo Lombardia owner representative Davide Pacca, Regione Lombardia (right) and architect Henry Cobb, Pei Cobb Freed & Partners (middle).

Top Left: Detail view of the 2012 Innovation Award-winning façade at Al Bahar Towers.

Bottom Left: CTBUH Awards Jury Chair Richard Cook (far left) presents the Innovation Award to the Al Bahar Towers team (from left to right): architect Peter Oborn, Aedas; owner representative Omar Liaqat, Abu Dhabi Investment Council; and engineer Peter Chipchase, Arup.

Top Right: The new Dean of Architecture at IIT, Wiel Arets (far left) presents the overall Best Tall Building Worldwide award to the Doha Tower team (from left to right): owner representative Hassan Al Duhaimi, HBS Company; architect Hafid Rakem, Ateliers Jean Nouvel; and engineer Peter Terrell, Terrell Group.

Bottom Right: Doha Tower, Doha, 2012 Best Building Middle East & Africa, and overall Best Tall Building Worldwide winner.

Overview of All Past Winners

Previous Lifetime Achievement Award Recipients:

2002:

Lynn S. Beedle Award
Dr. Lynn S. Beedle

2003:

Lynn S. Beedle Award
Charles DeBenedittis

2004:

Lynn S. Beedle Award
Gerald Hines

Fazlur R. Khan Medal
Leslie E. Robertson

2005:

Lynn S. Beedle Award
Dr. Alan Davenport

Fazlur R. Khan Medal
Dr. Werner Sobek

2006:

Lynn S. Beedle Award
Dr. Ken Yeang

Fazlur R. Khan Medal
Srinivasa "Hal" Iyengar

2007:

Lynn S. Beedle Award
Lord Norman Foster

Fazlur R. Khan Medal
Dr. Farzad Naeim

2008:

Lynn S. Beedle Award
Cesar Pelli

Fazlur R. Khan Medal
William F. Baker

2009:

Lynn S. Beedle Award
John C. Portman, Jr.

Fazlur R. Khan Medal
Dr. Prabodh V. Banavalkar

2010:

Lynn S. Beedle Award
William Pedersen

Fazlur R. Khan Medal
Ysrael A. Seinuk

2011:

Lynn S. Beedle Award
Adrian Smith

Fazlur R. Khan Medal
Dr. Akira Wada

2012:

Lynn S. Beedle Award
Helmut Jahn

Fazlur R. Khan Medal
Charles Thornton

Fazlur R. Khan Medal
Richard Tomasetti

Previous Best Tall Building Award Recipients:

2007:

Best Tall Building – Overall
Beetham Tower,
Manchester, UK

Best Sustainable Tall Building
Hearst Tower,
New York City, USA

2008:

Best Tall Building – Overall & Best Tall Building – Asia & Australasia
Shanghai WFC,
Shanghai, China

Best Tall Building – Americas
The New York Times Building,
New York City, USA

Best Tall Building – Europe
51 Lime Street,
London, UK

Best Tall Building – Middle East & Africa
Bahrain World Trade Center,
Manama, Bahrain

2009:

Best Tall Building – Overall & Best Tall Building – Asia & Australasia
Linked Hybrid,
Beijing, China

Best Tall Building – Americas
Manitoba Hydro Place,
Winnipeg, Canada

Best Tall Building – Europe
The Broadgate Tower,
London, UK

Best Tall Building – Middle East & Africa
Tornado Tower,
Doha, Qatar

2010:

"Global Icon" Award & Best Tall Building – Middle East & Africa
Burj Khalifa,
Dubai, UAE

Best Tall Building – Overall & Best Tall Building – Europe
Broadcasting Place,
Leeds, UK

2011:

Best Tall Building – Americas
Bank of America Tower,
New York City, USA

Best Tall Building – Asia & Australasia
Pinnacle @ Duxton,
Singapore

Best Tall Building – Overall & Best Tall Building – Europe
KfW Westerkade,
Frankfurt, Germany

Best Tall Building – Americas
Eight Spruce Street,
New York City, USA

Best Tall Building – Asia & Australasia
Guangzhou International Finance Center,
Guangzhou, China

Best Tall Building – Middle East & Africa
The Index,
Dubai, UAE

2012:

Best Tall Building – Overall & Best Tall Building – Middle East & Africa
Doha Tower,
Doha, Qatar

Best Tall Building – Americas
Absolute Towers,
Mississauga, Canada

Best Tall Building – Asia & Australasia
1 Bligh Street, *Sydney, Australia*

Best Tall Building – Europe
Palazzo Lombardia,
Milan, Italy

Tall Building Innovation Award
Al Bahar Towers,
Abu Dhabi, UAE

CTBUH Height Criteria

The Council on Tall Buildings and Urban Habitat is the official arbiter of the criteria upon which tall building height is measured, and the title of "The World's (or Country's, or City's) Tallest Building" determined. The Council maintains an extensive set of definitions and criteria for measuring and classifying tall buildings which are the basis for the official "100 Tallest Buildings in the World" list (see pages 211–215).

What is a Tall Building?
There is no absolute definition of what constitutes a "tall building." It is a building that exhibits some element of "tallness" in one or more of the following categories:

a) Height Relative to Context
It is not just about height, but about the context in which it exists. Thus, whereas a 14-story building may not be considered a tall building in a high-rise city such as Chicago or Hong Kong, in a provincial European city or a suburb this may be distinctly taller than the urban norm.

b) Proportion
Again, a tall building is not just about height, but also about proportion. There are numerous buildings which are not particularly high, but are slender enough to give the appearance of a tall building, especially against low urban backgrounds. Conversely, there are numerous big/large-footprint buildings which are quite tall, but their size/floor area rules them out as being classed as a tall building.

c) Tall Building Technologies
If a building contains technologies which may be attributed as being a product of "tall" (e.g., specific vertical transport technologies, structural wind bracing as a product of height, etc.), then this building can be classed as a tall building.

Although number of floors is a poor indicator of defining a tall building due to the changing floor-to-floor height between differing buildings and functions (e.g., office versus residential usage), a building of 14 or more stories – or over 50 meters (165 feet) in height – could perhaps be used as a threshold for considering it a "tall building."

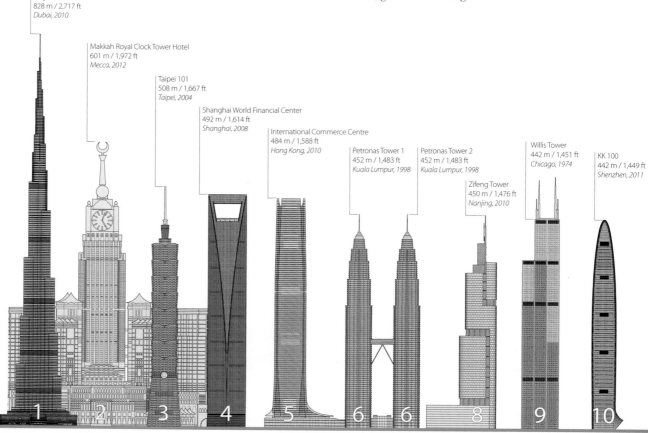

Above: Diagram of the World's Tallest 20 Buildings according to the CTBUH Height Criteria of "Height to Architectural Top" (as of July 2013)

What are Supertall and Megatall Buildings?

The CTBUH defines "supertall" as a building over 300 meters (984 feet) in height, and a "megatall" as a building over 600 meters (1,968 feet) in height. Although great heights are now being achieved with built tall buildings – in excess of 800 meters (2,600 feet) – as of July 2013 there are only 73 supertall and 2 megatall buildings completed and occupied globally. Thus the completion of a supertall building is still a significant milestone.

How is a Tall Building Measured?

The CTBUH recognizes tall building height in three categories:

1. Height to Architectural Top: Height is measured from the level[1] of the lowest, significant,[2] open-air,[3] pedestrian[4] entrance to the architectural top of the building, including spires, but not including antennae, signage, flagpoles, or other functional-technical equipment.[5] This measurement is the most widely utilized and is employed to define the Council on Tall Buildings and Urban Habitat (CTBUH) rankings of the "World's Tallest Buildings."

2. Highest Occupied Floor: Height is measured from the level[1] of the lowest, significant,[2] open-air,[3] pedestrian[4] entrance to the finished floor level of the highest occupied[6] floor within the building.

3. Height to Tip: Height is measured from the level[1] of the lowest, significant,[2] open-air,[3] pedestrian[4] entrance to the highest point of the building, irrespective of material or function of the highest element (i.e., including antennae, flagpoles, signage, and other functional-technical equipment).

Number of Floors:

The number of floors should include the ground floor level and be the number of main floors above ground, including any significant mezzanine floors and major mechanical plant floors. Mechanical mezzanines should not be included if they have a significantly smaller floor area than the major floors below. Similarly, mechanical penthouses or plant rooms protruding above the general roof area should not be counted. Note: CTBUH floor counts may differ from published accounts, as it is common in some regions of the world for certain floor levels not to be included (e.g., the level 4, 14, 24, etc. in Hong Kong).

Building Usage:

What is the difference between a tall building and a telecommunications/observation tower?

A tall "building" can be classed as such (as opposed to a telecommunications/observation tower) and is eligible for the "tallest" lists if at least 50 percent of its height is occupied by usable floor area.

Single-Function and Mixed-Use Buildings:

A **single-function** tall building is defined as one where 85 percent or more of its total floor area is dedicated to a single usage.

A **mixed-use** tall building contains two or more functions (or uses), where each of the functions occupy a significant proportion[7] of the tower's total space. Support areas such as car parks and mechanical plant space do not constitute mixed-use functions. Functions are denoted on CTBUH "Tallest" lists in descending order, e.g., "hotel/office" indicates hotel function above office function.

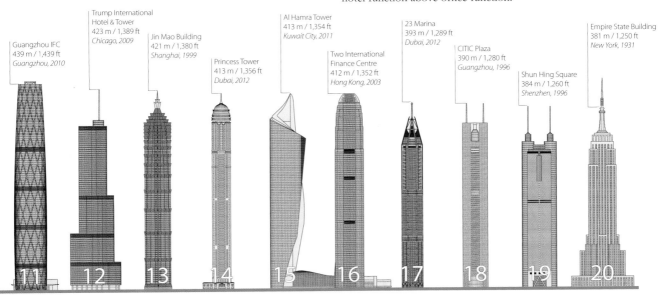

11. Guangzhou IFC — 439 m / 1,439 ft — *Guangzhou, 2010*
12. Trump International Hotel & Tower — 423 m / 1,389 ft — *Chicago, 2009*
13. Jin Mao Building — 421 m / 1,380 ft — *Shanghai, 1999*
14. Princess Tower — 413 m / 1,356 ft — *Dubai, 2012*
15. Al Hamra Tower — 413 m / 1,354 ft — *Kuwait City, 2011*
16. Two International Finance Centre — 412 m / 1,352 ft — *Hong Kong, 2003*
17. 23 Marina — 393 m / 1,289 ft — *Dubai, 2012*
18. CITIC Plaza — 390 m / 1,280 ft — *Guangzhou, 1996*
19. Shun Hing Square — 384 m / 1,260 ft — *Shenzhen, 1996*
20. Empire State Building — 381 m / 1,250 ft — *New York, 1931*

Building Status:

Complete (Completion)
A building is considered to be "complete" (and added to the CTBUH Tallest Buildings lists) if it fulfills all of the following three criteria: (i) topped out structurally and architecturally, (ii) fully clad, and (iii) open for business, or at least partially occupiable.

Under Construction (Start of Construction)
A building is considered to be "Under Construction" once site clearing has been completed and foundation/piling work has begun.

Topped Out
A building is considered to be "Topped Out" when it is under construction, and has reached its full height both structurally and architecturally (e.g., including its spires, parapets, etc.).

Proposed (Proposal)
A building is considered to be "proposed" (i.e., a real proposal) when it fulfills all of the following criteria: (i) has a specific site with ownership interests within the building development team, (ii) has a full professional design team progressing the design beyond the conceptual stage, (iii) has obtained, or is in the process of obtaining, formal planning consent/legal permission for construction, and (iv) has a full intention to progress the building to construction and completion.

Vision
A building is considered to be a "vision" when it either: (i) is in the early stages of inception and does not yet fulfill the criteria under the "proposal" category, (ii) was a proposal that never advanced to the construction stages, or (iii) was a theoretical proposition.

Demolished
A building is considered to be "demolished" after it has been destroyed by controlled end-of-life demolition, fire, natural catastrophe, war, terrorist attack, or through other means intended or unintended.

Structural Material:
A **steel** tall building is defined as one where the main vertical and lateral structural elements and floor systems are constructed from steel.

A **concrete** tall building is defined as one where the main vertical and lateral structural elements and floor systems are constructed from concrete.

A **composite** tall building utilizes a combination of both steel and concrete acting compositely in the main structural elements, thus including a steel building with a concrete core.

A **mixed-structure** tall building is any building that utilizes distinct steel or concrete systems above or below each other. There are two main types of mixed structural systems: a **steel/concrete** tall building indicates a steel structural system located above a concrete structural system, with the opposite true of a **concrete/steel** building.

Additional Notes on Structural Material:
(i) If a tall building is of steel construction with a floor system of concrete planks on steel beams, it is considered a steel tall building.
(ii) If a tall building is of steel construction with a floor system of a concrete slab on steel beams, it is considered a steel tall building.
(iii) If a tall building has steel columns plus a floor system of concrete beams, it is considered a composite tall building.

Footnotes:

[1] Level: finished floor level at threshold of the lowest entrance door.

[2] Significant: the entrance should be predominantly above existing or pre-existing grade and permit access to one or more primary uses in the building via elevators, as opposed to ground floor retail or other uses which solely relate/connect to the immediately adjacent external environment. Thus entrances via below-grade sunken plazas or similar are not generally recognized. Also note that access to car park and/or ancillary/support areas are not considered significant entrances.

[3] Open-air: the entrance must be located directly off of an external space at that level that is open to air.

[4] Pedestrian: refers to common building users or occupants and is intended to exclude service, ancillary, or similar areas.

[5] Functional-technical equipment: this is intended to recognize that functional-technical equipment is subject to removal/addition/change as per prevalent technologies, as is often seen in tall buildings (e.g., antennae, signage, wind turbines, etc. are periodically added, shortened, lengthened, removed, and/or replaced).

[6] Highest occupied floor: this is intended to recognize conditioned space which is designed to be safely and legally occupied by residents, workers, or other building users on a consistent basis. It does not include service or mechanical areas which experience occasional maintenance access, etc.

[7] This "significant proportion" can be judged as 15 percent or greater of either: (i) the total floor area, or (ii) the total building height, in terms of number of floors occupied for the function. However, care should be taken in the case of supertall towers. For example a 20-story hotel function as part of a 150-story tower does not comply with the 15 percent rule, though this would clearly constitute mixed-use.

100 Tallest Buildings in the World (as of July 2013)

The Council maintains the official list of the 100 Tallest Buildings in the World, which are ranked based on the height to architectural top, and includes not only completed buildings, but also buildings currently under construction. However, a building does not receive an official ranking number until it is completed.

Color Key:
Buildings in black are completed and officially ranked.
Buildings in blue are under construction and have topped out.
Buildings in green are under construction, but have not yet topped out.

Rank	Building Name	City	Height (meters)	Height (feet)	Stories	Year	Material	Function
	Kingdom Tower	Jeddah	1,000+*	3,281+*	167	2019	concrete	residential / hotel / office
1	Burj Khalifa	Dubai	828	2,717	163	2010	steel / concrete	office / residential / hotel
	Ping An Finance Center	Shenzhen	660	2,165	115	2016	composite	office
	Wuhan Greenland Center	Wuhan	636	2,087	125	2017	composite	hotel / residential / office
	Shanghai Tower	Shanghai	632	2,073	128	2014	composite	hotel / office
2	Makkah Royal Clock Tower Hotel	Mecca	601	1,972	120	2012	steel / concrete	other / hotel / multiple
	Goldin Finance 117	Tianjin	597	1,957	128	2016	composite	hotel / office
	Lotte World Tower	Seoul	555	1,819	123	2015	composite	hotel / office
	One World Trade Center	New York City	541**	1,776**	104	2014	composite	office
	The CTF Guangzhou	Guangzhou	530	1,739	111	2017	composite	hotel / residential / office
	Tianjin Chow Tai Fook Binhai Center	Tianjin	530	1,739	97	2016	composite	residential / hotel / office
	Zhongguo Zun	Beijing	528	1,732	108	2018	composite	office
	Busan Lotte Town Tower	Busan	510	1,674	107	2016	composite	residential / hotel / office
3	Taipei 101	Taipei	508	1,667	101	2004	composite	office
4	Shanghai World Financial Center	Shanghai	492	1,614	101	2008	composite	hotel / office
5	International Commerce Centre	Hong Kong	484	1,588	108	2010	composite	hotel / office
	International Commerce Center 1	Chongqing	468	1,535	99	2016	composite	hotel / office
	Tianjin R&F Guangdong Tower	Tianjin	468	1,535	91	2016	composite	residential / hotel / office
	Lakhta Center	St. Petersburg	463	1,517	86	2018	composite	office
	Riverview Plaza A1	Wuhan	460	1,509	82	2016	–	hotel / office
6	Petronas Tower 1	Kuala Lumpur	452	1,483	88	1998	composite	office
6	Petronas Tower 2	Kuala Lumpur	452	1,483	88	1998	composite	office
	Suzhou Supertower	Suzhou	452	1,483	92	2016	–	residential / hotel / office
8	Zifeng Tower	Nanjing	450	1,476	66	2010	composite	hotel / office
9	Willis Tower	Chicago	442	1,451	108	1974	steel	office
	World One	Mumbai	442	1,450	117	2015	composite	residential
10	KK100	Shenzhen	442	1,449	100	2011	composite	hotel / office
11	Guangzhou International Finance Center	Guangzhou	439	1,439	103	2010	composite	hotel / office
	Wuhan Center	Wuhan	438	1,437	88	2015	composite	hotel / residential / office
	Diamond Tower	Jeddah	432	1,417	93	–	–	residential
	Marina 101	Dubai	432	1,417	101	2014	concrete	residential / hotel
	432 Park Avenue	New York City	426	1,397	85	2015	concrete	residential
12	Trump International Hotel & Tower	Chicago	423	1,389	98	2009	concrete	residential / hotel
13	Jin Mao Tower	Shanghai	421	1,380	88	1999	composite	hotel / office
14	Princess Tower	Dubai	413	1,356	101	2012	steel / concrete	residential
15	Al Hamra Tower	Kuwait City	413	1,354	80	2011	concrete	office
16	Two International Finance Centre	Hong Kong	412	1,352	88	2003	composite	office
	Huaguoyuan Tower 1	Guiyang	406	1,332	64	–	–	–
	Huaguoyuan Tower 2	Guiyang	406	1,332	64	–	–	–
	Nanjing Olympic Suning Tower	Nanjing	400	1,312	89	2016	–	residential / hotel / office

* estimated height
** height of One World Trade Center has not yet been ratified by the CTBUH Height Committee

Rank	Building Name	City	Height (meters)	Height (feet)	Stories	Year	Material	Function
	Ningbo Center	Ningbo	398*	1,306*	–	2017	–	hotel / residential / office
17	23 Marina	Dubai	393	1,289	90	2012	concrete	residential
18	CITIC Plaza	Guangzhou	390	1,280	80	1996	concrete	office
	Logan Century Center 1	Nanning	386	1,266	82	2017	–	hotel / office
	Capital Market Authority Headquarters	Riyadh	385	1,263	77	2014	composite	office
19	Shun Hing Square	Shenzhen	384	1,260	69	1996	composite	office
	Eton Place Dalian Tower 1	Dalian	383	1,257	80	2014	composite	hotel / office
	Abu Dhabi Plaza	Astana	382	1,253	88	2017		residential
	World Trade Center Abu Dhabi – The Residences	Abu Dhabi	381	1,251	88	2013	concrete	residential
20	Empire State Building	New York City	381	1,250	102	1931	steel	office
21	Elite Residence	Dubai	380	1,248	87	2012	concrete	residential
	Gemdale Gangxia Tower 1	Shenzhen	375	1,230	–	2016	–	residential / office
22	Central Plaza	Hong Kong	374	1,227	78	1992	concrete	office
	Oberoi Oasis Tower B	Mumbai	372	1,220	82	2015	concrete	residential
	The Address The BLVD	Dubai	370	1,214	72	2015		residential / hotel
23	Bank of China Tower	Hong Kong	367	1,205	72	1990	composite	office
24	Bank of America Tower	New York City	366	1,200	55	2009	composite	office
	Dalian International Trade Center	Dalian	365	1,199	86	2015	composite	residential / office
	VietinBank Business Center Office Tower	Hanoi	363	1,191	68	2016	composite	office
25	Almas Tower	Dubai	360	1,181	68	2008	concrete	office
25	The Pinnacle	Guangzhou	360	1,181	60	2012	concrete	office
	Federation Towers – Vostok Tower	Moscow	360	1,181	93	2015	concrete	residential / hotel / office
27	JW Marriott Marquis Hotel Dubai Tower 1	Dubai	355	1,166	82	2012	concrete	hotel
27	JW Marriott Marquis Hotel Dubai Tower 2	Dubai	355	1,166	82	2013	concrete	hotel
29	Emirates Tower One	Dubai	355	1,163	54	2000	composite	office
	Forum 66 Tower 2	Shenyang	351	1,150	68	2015	composite	office
30	Tuntex Sky Tower	Kaohsiung	348	1,140	85	1997	composite	hotel / office
31	Aon Center	Chicago	346	1,136	83	1973	steel	office
32	The Center	Hong Kong	346	1,135	73	1998	steel	office
33	John Hancock Center	Chicago	344	1,128	100	1969	steel	residential / office
	Four Seasons Place	Kuala Lumpur	343	1,124	65	2017	–	residential / hotel
	ADNOC Headquarters	Abu Dhabi	342	1,122	76	2014	concrete	office
	Ahmed Abdul Rahim Al Attar Tower	Dubai	342	1,122	76	2014	concrete	residential
	Xiamen International Centre	Xiamen	340	1,115	61	2016	composite	office
	The Wharf Times Square 1	Wuxi	339	1,112	68	2015	composite	hotel / residential
	Chongqing World Financial Center	Chongqing	339	1,112	73	2014	composite	office
	Mercury City Tower	Moscow	339	1,112	75	2013	concrete	residential / office
	Four Seasons Tower	Tianjin	338	1,109	65	2015	composite	residential / hotel
	Orchid Crown Tower A	Mumbai	337	1,106	75	2015	concrete	residential
	Orchid Crown Tower B	Mumbai	337	1,106	75	2015	concrete	residential
34	Tianjin Global Financial Center	Tianjin	337	1,105	75	2011	composite	office
34	The Torch	Dubai	337	1,105	79	2011	concrete	residential
36	Keangnam Hanoi Landmark Tower	Hanoi	336	1,102	72	2012	concrete	hotel / residential / office
	Oko Tower 1	Moscow	336	1,101	91	2015	concrete	residential / hotel
	DAMAC Residenze	Dubai	335	1,099	86	2016	steel / concrete	residential
37	Shimao International Plaza	Shanghai	333	1,094	60	2006	concrete	hotel / office
	China Chuneng Tower	Shenzhen	333	1,093	–	2016	–	–
38	Rose Rayhaan by Rotana	Dubai	333	1,093	71	2007	composite	hotel
	Tianjin Kerry Center	Tianjin	333	1,093	72	2016	steel	office
	Modern Media Center	Changzhou	332	1,089	57	2013	composite	office

* estimated height

Rank	Building Name	City	Height (meters)	Height (feet)	Stories	Year	Material	Function
39	Minsheng Bank Building	Wuhan	331	1,086	68	2008	steel	office
40	China World Tower	Beijing	330	1,083	74	2010	composite	hotel / office
	Ryugyong Hotel	Pyongyang	330	1,083	105	–	concrete	hotel / office
	Gate of Kuwait Tower	Kuwait City	330	1,083	80	2016	concrete	hotel / office
	Wuhan Qiakou Project 1	Wuhan	330	1,083	66	2016	–	office
	The Skyscraper	Dubai	330	1,083	66	–	–	office
	Suning Plaza Tower 1	Zhenjiang	330	1,082	77	2016	–	–
	Hon Kwok City Center	Shenzhen	329	1,081	80	2015	composite	residential / office
	Concord International Centre	Chongqing	328	1,076	62	2016	composite	hotel / office
	Wuxi Suning Plaza 1	Wuxi	328	1,076	68	2014	composite	hotel / office
	Nanjing World Trade Center Tower 1	Nanjing	328	1,076	69	2016	composite	hotel / office
41	Longxi International Hotel	Jiangyin	328	1,076	72	2011	composite	residential / hotel
	Al Yaqoub Tower	Dubai	328	1,076	69	2013	concrete	residential / hotel
	Greenland Center Tower 1	Qingdao	327	1,074	74	–	–	hotel / office
42	The Index	Dubai	326	1,070	80	2010	concrete	residential / office
43	Deji Plaza Phase 2	Nanjing	324	1,063	62	2013	composite	office
43	The Landmark	Abu Dhabi	324	1,063	72	2013	concrete	residential / office
	Yantai Shimao No. 1 The Harbour	Yantai	323	1,060	59	2014	composite	residential / hotel / office
45	Q1 Tower	Gold Coast	323	1,058	78	2005	concrete	residential
46	Wenzhou Trade Center	Wenzhou	322	1,056	68	2011	concrete	hotel / office
	Lamar Tower 1	Jeddah	322	1,056	70	2015	concrete	residential / office
47	Burj Al Arab	Dubai	321	1,053	56	1999	composite	hotel
48	Nina Tower	Hong Kong	320	1,051	80	2006	concrete	hotel / office
	Palais Royale	Mumbai	320	1,050	88	2014	concrete	residential
	Chongqing IFS T1	Chongqing	320	1,048	64	2016	–	–
	White Magnolia Plaza 1	Shanghai	320	1,048	66	2015	composite	office
	Zhujiang New City Tower	Guangzhou	319	1,046	67	2015	composite	office
49	Chrysler Building	New York City	319	1,046	77	1930	steel	office
49	New York Times Tower	New York City	319	1,046	52	2007	steel	office
	Jiuzhou International Tower	Nanning	318	1,043	71	2015	–	–
	Riverside Century Plaza Main Tower	Wuhu	318	1,043	66	2015	composite	hotel / office
	Runhua Global Center 1	Changzhou	318	1,043	72	2015	composite	office
	United International Mansion	Chongqing	318	1,043	67	2013	concrete	office
51	HHHR Tower	Dubai	318	1,042	72	2010	concrete	residential
	Bashang Jie North Tower	Hefei	317	1,040	–	2015	–	–
52	Bank of America Plaza	Atlanta	317	1,040	55	1993	composite	office
	Youth Olympics Center Tower 1	Nanjing	315	1,032	68	2015	composite	–
	Maha Nakhon	Bangkok	313	1,028	77	2015	concrete	residential / hotel
	Yunrun International Tower	Huaiyin	312	1,024	75	2015	–	office
	The Stratford Residences	Makati	312	1,024	74	2015	concrete	residential
	Moi Center Tower A	Shenyang	311	1,020	75	2013	composite	hotel / office
53	U.S. Bank Tower	Los Angeles	310	1,018	73	1990	steel	office
54	Menara Telekom	Kuala Lumpur	310	1,017	55	2001	concrete	office
54	Ocean Heights	Dubai	310	1,017	83	2010	concrete	residential
	Bashang Jie South Tower	Hefei	310	1,016	–	2015	–	–
56	Pearl River Tower	Guangzhou	309	1,015	71	2012	composite	office
	Guangzhou Fortune Center	Guangzhou	309	1,015	73	2015	composite	office
57	Emirates Tower Two	Dubai	309	1,014	56	2000	concrete	hotel
	Eurasia	Moscow	309	1,013	72	2014	composite	hotel / office
	Guangfa Securities Headquarters	Guangzhou	308	1,010	62	2016	–	office

Rank	Building Name	City	Height (meters)	Height (feet)	Stories	Year	Material	Function
	Burj Rafal	Riyadh	308	1,010	68	2014	concrete	residential / hotel
	Wanda Plaza 1	Kunming	307	1,008	67	2016	composite	office
	Wanda Plaza 2	Kunming	307	1,008	67	2016	composite	office
58	Cayan Tower	Dubai	307	1,008	73	2013	concrete	residential
59	Franklin Center – North Tower	Chicago	307	1,007	60	1989	composite	office
	Lokhandwala Minerva	Mumbai	307	1,007	83	2015	concrete	residential
	East Pacific Center Tower A	Shenzhen	306	1,004	85	2013	concrete	residential
	One57	New York City	306	1,004	79	2014	steel / concrete	residential / hotel
60	The Shard	London	306	1,004	73	2013	composite	residential / hotel / office
61	JPMorgan Chase Tower	Houston	305	1,002	75	1982	composite	office
61	Etihad Towers T2	Abu Dhabi	305	1,002	80	2011	concrete	residential
63	Northeast Asia Trade Tower	Incheon	305	1,001	68	2011	composite	residential / hotel / office
	Wuxi Maoye City – Marriott Hotel	Wuxi	304	997	68	2013	composite	hotel
64	Baiyoke Tower II	Bangkok	304	997	85	1997	concrete	hotel
	Shenzhen World Finance Center	Shenzhen	304	997	68	2016	composite	office
65	Two Prudential Plaza	Chicago	303	995	64	1990	concrete	office
	KAFD World Trade Center	Riyadh	303	994	67	2014	concrete	office
	Diwang International Fortune Center	Liuzhou	303	994	75	2014	composite	residential / hotel / office
66	Leatop Plaza	Guangzhou	303	993	64	2012	composite	office
67	Wells Fargo Plaza	Houston	302	992	71	1983	steel	office
67	Kingdom Centre	Riyadh	302	992	41	2002	steel / concrete	residential / hotel / office
69	The Address	Dubai	302	991	63	2008	concrete	residential / hotel
70	Capital City Moscow Tower	Moscow	302	990	76	2010	concrete	residential
	Gate of the Orient	Suzhou	302	990	68	2014	composite	residential / hotel / office
	Heung Kong Tower	Shenzhen	301	987	70	2014	composite	hotel / office
	Dubai Pearl Tower East	Dubai	300	984	73	2017	concrete	—
	Dubai Pearl Tower South	Dubai	300	984	73	2017	concrete	—
	Dubai Pearl Tower North	Dubai	300	984	73	2017	concrete	—
	Riverfront Times Square	Shenzhen	300	984	64	2016	composite	hotel / office
71	Aspire Tower	Doha	300	984	36	2007	composite	hotel / office
	Namaste Tower	Mumbai	300*	984*	62	2015	concrete	hotel / office
	Jin Wan Plaza 1	Tianjin	300	984	66	2015	—	hotel / office
	Abeno Harukas	Osaka	300	984	62	2014	steel	hotel / office / retail
	Greenland Center Tower 1	Zhengzhou	300	984	78	2016	composite	office
	Greenland Center Tower 2	Zhengzhou	300	984	78	2016	composite	office
71	Arraya Tower	Kuwait City	300	984	60	2009	concrete	office
	Torre Costanera	Santiago	300	984	64	2013	concrete	office
	NBK Tower	Kuwait City	300	984	70	2014	concrete	office
	Shenglong Global Center	Fuzhou	300	984	57	2016	—	office
71	Doosan Haeundae We've the Zenith Tower A	Busan	300	984	80	2011	concrete	residential
	Orchid Crown Tower C	Mumbai	300*	984*	75	2015	concrete	residential
	Dubai Pearl Tower West	Dubai	300	984	73	2017	concrete	residential
	Supernova	Noida	300	984	80	2015	—	residential
	Langham Hotel Tower	Dalian	300	983	74	2015	—	residential / hotel
74	One Island East	Hong Kong	298	978	68	2008	concrete	office
74	First Bank Tower	Toronto	298	978	72	1975	steel	office
	Yujiabao Administrative Services Center	Tianjin	298	978	60	2015	—	office
	Ilham Baru Tower	Kuala Lumpur	298	978	64	2015	concrete	residential / office
	Four World Trade Center	New York City	298	977	64	2013	composite	office
76	Eureka Tower	Melbourne	297	975	91	2006	concrete	residential

* estimated height

Rank	Building Name	City	Height (meters)	Height (feet)	Stories	Year	Material	Function
	Dacheng Financial Business Center Tower A	Kunming	297	974	–	2015	steel	hotel / office
77	Comcast Center	Philadelphia	297	974	57	2008	composite	office
78	Landmark Tower	Yokohama	296	972	73	1993	steel	hotel / office
	Park Hyatt Guangzhou	Guangzhou	296	972	66	2013	composite	residential / hotel / office
79	Emirates Crown	Dubai	296	971	63	2008	concrete	residential
	Xiamen Shimao Cross-Strait Plaza Tower B	Xiamen	295	969	67	2015	–	office
80	Khalid Al Attar Tower 2	Dubai	294	965	66	2011	concrete	hotel
81	311 South Wacker Drive	Chicago	293	961	65	1990	concrete	office
	Lamar Tower 2	Jeddah	293	961	62	2015	concrete	residential / office
	Greenland Puli Center	Jinan	293	960	61	2015	composite	residential / office
82	Sky Tower	Abu Dhabi	292	959	74	2010	concrete	residential / office
83	Haeundae I Park Marina Tower 2	Busan	292	958	72	2011	composite	residential
84	SEG Plaza	Shenzhen	292	957	71	2000	concrete	hotel / office
	Indiabulls Sky Suites	Mumbai	291	955	75	2015	concrete	residential
85	70 Pine Street	New York City	290	952	67	1932	steel	office
	Powerlong Center Tower 1	Tianjin	290	951	59	2014	composite	office
	Hunter Douglas International Plaza	Guiyang	290	951	69	2014	composite	office
	Tanjong Pagar Centre	Singapore	290	951	68	2016	–	residential / hotel / office
	Dongguan TBA Tower	Dongguan	289	948	68	2013	composite	hotel / office
	Jiangxi Nanchang Greenland Central Plaza 1	Nanchang	289	948	59	2014	composite	office
	Jiangxi Nanchang Greenland Central Plaza 2	Nanchang	289	948	59	2014	composite	office
	Busan International Finance Center Landmark Tower	Busan	289	948	63	2014	–	office
86	Key Tower	Cleveland	289	947	57	1991	composite	office
87	Shaoxing Shimao Crown Plaza	Shaoxing	288	946	60	2012	composite	hotel / office
	Soochow International Plaza East Tower	Huzhou	288	945	–	2014	composite	hotel / office
	Kaisa Center	Huizhou	288	945	66	2014	composite	hotel / office
88	Yingli International Finance Centre	Chongqing	288	945	58	2012	concrete	office
88	Plaza 66	Shanghai	288	945	66	2001	concrete	office
88	One Liberty Place	Philadelphia	288	945	61	1987	steel	office
	Soochow International Plaza West Tower	Huzhou	288	945	–	2014	composite	residential
	Chongqing Poly Tower	Chongqing	287	941	58	2013	concrete	office / hotel
91	Sulafa Tower	Dubai	285	935	75	2010	concrete	residential
91	Millennium Tower	Dubai	285	935	59	2006	concrete	residential
93	Tomorrow Square	Shanghai	285	934	60	2003	composite	residential / hotel / office
94	Columbia Center	Seattle	284	933	76	1984	composite	office
95	Three International Finance Center	Seoul	284	932	55	2012	composite	office
	D1 Tower	Dubai	284	932	80	2014	concrete	residential
95	Trump Ocean Club International Hotel & Tower	Panama City	284	932	70	2011	concrete	residential / hotel
97	Chongqing World Trade Center	Chongqing	283	929	60	2005	concrete	office
98	Cheung Kong Centre	Hong Kong	283	928	63	1999	steel	office
99	The Trump Building	New York City	283	927	71	1930	steel	office
	City of Lights C1 Tower	Abu Dhabi	282	926	62	2014	concrete	office
100	Suzhou RunHua Global Building A	Suzhou	282	925	49	2010	composite	office

Index of Buildings

1214 Fifth Avenue, *New York;* 42
30 St Mary Axe, *London;* 158
6 Remez Tower, *Tel Aviv;* 144

ADAC Headquarters, *Munich;* 118
Alamanda Office Tower, *Jakarta;* 104
Ann & Robert H. Lurie Children's Hospital, *Chicago;* 44
ARK Hills Sengokuyama Mori Tower, *Tokyo;* 104

Bow, The, *Calgary;* 28
Brookfield Place, *Perth;* 80

C&D International Tower, *Xiamen;* 64
CCTV Headquarters, *Beijing;* 58
City Tower Kobe Sannomiya, *Kobe;* 105
Coast at Lakeshore East, *Chicago;* 52

Devon Energy Center, *Oklahoma City;* 34
Diplomat Commercial Office Tower, *Manama;* 154
Dolphin Plaza, *Hanoi;* 105

Ellipse 360, The, *New Taipei City;* 100

Faire Tower, *Ramat-Gan;* 154
Frishman 46, *Tel Aviv;* 155

Gate Towers, *Abu Dhabi;* 148

Hangzhou Civic Center, *Hangzhou;* 82
Helicon, *San Pedro Garza Garcia;* 52
Huarun Tower, *Chengdu;* 106
Hysan Place, *Hong Kong;* 84

I Tower, *Incheon;* 88
International Finance Centre, *Seoul;* 86

Japan Post Tower, *Tokyo;* 90
JW Marriott Marquis, *Dubai;* 152

KONE UltraRope; 168

LOVFT, *Santa Catarina;* 46

Mercedes House, *New York;* 48
Mercury City, *Moscow;* 130

Nakanoshima Festival Tower, *Osaka – Megatruss Seismic Isolation Structure;* 172
NBF Osaki Building, *Tokyo;* 92
Net Metropolis, *Manila;* 106
New Babylon, *The Hague;* 122
No. 1 Great Marlborough Street, *Manchester;* 134

Pacifica Honolulu, *Honolulu;* 53
PARKROYAL on Pickering, *Singapore;* 68
Pearl River Tower, *Guangzhou;* 72
Poly Corporation Headquarters, *Beijing – Rocker Façade Support System;* 180
Pyne, *Bangkok;* 107

Reflection Jomtien Beach, *Pattaya;* 107
Reforma 342, *Mexico City;* 53
Rush University Medical Center Hospital Tower, *Chicago;* 54

Shard, The, *London;* 112
Shenzhen Kerry Plaza Phase II, *Shenzhen;* 108
Shenzhen Stock Exchange, *Shenzhen;* 94
Shibuya Hikarie, *Tokyo;* 96
Sliced Porosity Block, *Chengdu;* 76
Soul, *Gold Coast;* 98
Sowwah Square, *Abu Dhabi;* 138

T30 Hotel, *Changsha – BSB Prefabricated Construction Method;* 164
Torre Begonias, *Lima;* 54

Torre Paseo Colón 1, *San José;* 55
Torre Unipol, *Bologna;* 134
Tour Total, *Berlin;* 126, 176
Tree House Residence Hall, *Boston;* 38
Trump International Hotel & Tower, *Toronto;* 50

Unicredit Tower, *Milan;* 132

Varyap Meridian Block A, *Istanbul;* 135

Yixing Dongjiu, *Yixing;* 108
Yokohama Mitsui Building, *Yokohama;* 109

Zhengzhou Greenland Plaza, *Zhengzhou;* 102

Index of Companies

20-12 Property Holdings, Inc.; 106
Ace-all; 88
ADAC; 119
Adamson Associates Architects; 113, 132
ADD Inc; 39
A. Dori Group; 154
Advance Mechanical Systems, Inc.; 52
AECOM; 106, 108
AESA Constructora; 54
Affiliated Engineers, Inc.; 44
AIG Global Real Estate; 86
AIG Korean Real Estate Development YH; 86
Aldar Properties PJSC; 149
Alejandro Molina; 54
Alex Meitlis Architecture & Design; 155
ALT Cladding & Design; 84
Altus Group; 29
Anderson Mikos Architects, Ltd.; 44
Ann & Robert H. Lurie Children's Hospital of Chicago; 44
Antonio Blanco; 54
Arabian Construction Company; 149
ARCADIS; 44
Archgroup Consultants; 152
Architects Hawaii, Ltd.; 53
Architectural Design & Research Institute of Zhejiang University; 82
ARGE Neubau; 119
Ariatta Ingengneria dei Sistemi, S.r.l.; 132
Arquitectonica; 54, 86, 149
Arquitectura y Diseño S.A.; 55
Arup; 59, 65, 77, 84, 90, 94, 106, 108, 113, 149, 159
Asahi-Dori 4-Chome Area Urban Redevelopment Union; 105
August Design Consultant Co., Ltd.; 107
Aurecon; 80
Baldridge & Associates Structural Engineering, Inc.; 53
Ballast Nedam NV en Boele & Van Eesteren; 123
Bar Akiva Engineers, Ltd.; 145, 155
Barkow Leibinger; 127, 177
Barr-Shoval Interior Design & Architecture; 154
Baum Architects; 86
BBM Müller-BBM; 127
BC&A; 108
BDSP Partnership Consulting Engineers; 159
BECA Carter Hollings & Ferner Pte Ltd.; 69
Beijing Jangho Curtain Wall Co., Ltd.; 102
Beijing Special Engineering Design and Research Institute; 181
Ben Avraham S. Engineers, Ltd.; 155
Benoy; 84
Betaprogetti; 134
BEYOND Landscape Design Group; 88
BG&E; 152
bKL Architecture LLC; 52
BLT Construction; 50
Bouwcombinatie New Babylon vof; 123
Brentwood Group; 134
BROAD Group; 165
Brookfield Multiplex; 50, 80, 152
Brookfield Office Properties; 80
Brook Van Dalen; 29
Bureau Veritas; 105
Buro Happold; 132, 135
bv Adviesburo T&H; 123
C3; 39
CADAE; 53
CA Immo Deutschland GmbH; 127
Callison, LLC; 106
Canaan-Shenhav Architects; 154
Canmoor; 134
CapitaLand China; 77
Carson McCulloch; 29
C&B Group; 98
CCW Associates Pte Ltd; 69
Cerami Associates; 29, 35

Chang Minwoo Structural Consultants; 86, 88
Chase Perdana Sdn Bhd; 154
China Academy of Building Research; 77
China Central Television; 59
China Construction Third Engineering Bureau Corp., Ltd.; 65, 77
China Resources Land Ltd.; 106
China State Construction Engineering Corporation; 59
Chomthai Design and Consultants Co., Ltd.; 105
Chung-Lin General Contractors, Ltd.; 100
Circuito S.A.; 55
CJSC Mercury City Tower; 130
Cladtech; 134
Claude Engle; 29
Colombo Costruzioni, S.p.A.; 132
Colonnier y Asociados, S. C.; 53
Columbus Engineering Co., Ltd.; 100
Consultancy Company of University of Civil Engineering; 105
Cooperativa Muratori e Braccianti di Carpi; 134
Corsmit Raadgevende Ingenieurs; 123
Cosentini Associates; 29, 35, 86
Council for Promotion of the Shibuya New Cultural District Development Project; 96
CR Construction; 106
CSAssociates, Inc.; 52
C|S Design Consultancy; 106
CTU Investments, Ltd.; 145
Cubica; 54
Cwg Danismanlik, Ltd.; 135
CYLA Design Associates; 44
CYMESA; 46, 52
Daewoo Engineering & Construction; 88
David Engineers, Ltd.; 145
Davis Langdon & Seah; 77, 84
dbHMS; 52
DBI Design; 98
Dennis Lau & Ng Chun Man; 84
Designcamp Moonpark dmp; 88
Devon Energy Corporation; 35
DHV Building and Industry; 59, 94
Diu Design Identity Unit Co., Ltd.; 107
DLC; 53
Do Architects; 145
DOCSA; 52
Dome Mimarlik; 135
Dongliang Light; 82
DP Architects Pte., Ltd.; 105
DPS; 46
Drees & Sommer; 127
Dreßler Bau GmbH; 177
DYPRO; 53
DYXY Architecture + Interiors; 104
ECADI; 59, 102, 108
EMF Griffiths; 98
Emirates Airlines; 152
Energydesign Asia; 100
energydesign Braunschweig GmbH; 127
Environmental Systems Design, Inc.; 54, 139
E.S.L–Eng.S.Lustig Consulting Engineers, Ltd.; 145
Ettinger Engineering Associates; 48
EURO-SAT Investments, Ltd.; 155
Evans Randall; 159
Exova Warrington; 149
Faire Fund; 154
First Construction Co., Ltd. of China Construction Third Engineering Bureau Corp., Ltd.; 108
Fitzpatrick + Partners; 80
FKN Fassadentechnik GmbH & Co. KG; 177
Flintco; 35
Fortress BV; 123
Fortune Consultants, Ltd.; 73, 102, 152
Foster + Partners; 29, 159
Frank Williams & Associates; 130

Friedman-kamar Engineers, Ltd.; 154
Front, Inc.; 53, 59
Fürstenau & Partner Ingenieurgesellschaft mbH; 127
Gammon Construction Limited; 84
Garza Maldonado; 53
General Construction Company of CCTEB; 94
General Office of CCTV New Site Construction & Development Program; 59
Gensler; 29, 35
George Sexton Lighting Design; 100
Glynn Tucker Engineers; 98
Goettsch Partners; 139
Gotham; 42
Gravity Green Ltd.; 65
Gravity Partnership Ltd.; 65
Greenland Group; 102
Green Star Builders, LLC; 48
Grocon; 98
Ground; 39
Grupo Inmobiliario del Parque S.A.; 55
GS Consortium; 86
GS Engineering & Construction; 86
Guangzhou Design Institute; 73
Guangzhou Pearl River Tower Properties Co., Ltd.; 73
Guangzhou Rongbaisheng Structural Design Firm; 106
GuD Planungsgesellschaft für Ingenieurbau mbH; 127, 177
Gurtz Electric Co.; 52
Gyungsung Architects & Engineers; 88
Haeahn Architrecure, Inc.; 88
Hangzhou Qianjiang New City Construction Headquarters; 82
Hann Tucker Associates; 139
Hanoi Construction Joint Stock Company No. 1; 105
HASSELL; 65, 80
Hawaiian Dredging Construction, Co.; 53
henke + rapolder Ingenieurgesellschaft; 119
HH Angus; 50
hhp Berlin Ingenieure für Brandschutz; 127
Hidi Rae Consulting; 50
Highrise Systems, Inc.; 73
Hill International, Ltd.; 149
Hilson Moran Partnership Ltd.; 159
Hines; 35, 132
Hitchcock Design Group; 54
Hodder+Partners; 134
Hoerr Schaudt; 54
Holder Construction; 35
H + R Reit; 29
HSG Zander; 130
Hysan Development Company Limited; 84
Ian Banham and Associates; 152
IBA; 54
IBC Engineering; 54
IBI Group, Inc.; 50
IDS Popov; 53
IECA International; 55
IIBYIV; 50
Ilshin Engineers & Construction; 88
Incheon Free Economic Zone Authority; 88
Infra Technology Service Co., Ltd.; 107
Inside/Outside; 59, 94
Integrated Environmental Solutions, Ltd.; 35, 139
International High-Rise Construction Centre, LLC; 130
Irie Miyake Architects & Engineers; 104
Iser Goldish Consulting, Ltd.; 145
Ismael Leyva Architects; 55
Ismail Khonji Associates; 154
Israel Berger & Associates; 42, 48
Israel David Engineers, Ltd.; 154
IVG; 159
Jacobs; 54
J&A Consultants S.n.c.; 132
JAHN; 90
James McHugh Construction Co.; 52
Jan Gehl Architects; 132
Japan Post Network Co., Ltd; 90
Jaros Baum & Bolles; 42

JCLI International; 106
J.E. Coulter Associates Ltd.; 50
Jenkins & Huntington, Inc.; 139
Jizhun Fangzhong Architectural Design Associates; 106
J.M. Lin Architect, P.C.; 100
Jules Wilson I.D.; 53
Juniper Development Group Pty., Ltd.; 98
Kajima Corporation; 92
Kaplan, Gehring, McCarroll Architectural Lighting; 102
Kardorff; 152
Kellam Berg; 29
Kenchiku Setsubi Sekkei Kenkyusho; 104
Kendall/Heaton Associates; 35
Kerry Development (Shenzhen) Co., Ltd.; 108
Khatib and Alami; 149
KJA; 29
Kohn Pedersen Fox Associates; 84
KONE Corporation, Ltd.; 169
Langan Engineering and Environmental Services; 48
Lassen Associates; 44
L&B Quantity Surveyors; 94
Leandro V. Locsin Partners; 106
Leber Rubes; 29
Ledcor Construction; 29, 53
Lerch Bates; 39, 44, 59
Leshem-Sheffer Environmental Quality, Ltd.; 154
Lewis Builds Corporation; 50
Lighting Design Alliance; 106
Lighting Planners Associates Pte Ltd.; 59, 69, 84
LJ Energy Pte Ltd.; 69
L'Observatoire International; 77
Loewenberg Architects LLC; 52
London Bridge Quarter, Ltd.; 113
LSD and Associates; 106
Luz en Arquitectura; 53
LWDesign; 152
Mace; 113
Mackie Consultants, LLC; 52
Magellan Development Group LLC; 52
Magnusson Klemencic Associates; 44
Major Development; 107
Majowiecki; 134
Martha Schwartz Partners, Ltd.; 139
Massachusetts State College Building Authority; 39
Matthews Southwest; 29
Meinhardt; 69, 84, 106, 149, 152
Mercury Development; 130
MFD Security Ltd.; 139
MF Ingenieros; 53
Mikyoung Kim Design; 44
Miranda & Nasi; 54
Mitsubishi Jisho Sekkei, Inc.; 90
Mitsui Fudosan Co., Ltd.; 109
M.M. Posokhin; 130
Mohamed Salahuddin Consulting Engineering Bureau; 154
Moises Farca; 53
Monolith Construction and Development Corporation; 106
Mori Building Co., Ltd.; 104
Mortenson; 44
Moshe Tzur Architects and Town Planners, Ltd.; 145
Mount Sinai Medical Center; 42
MSC Associati, S.r.l.; 132
Mubadala Real Estate & Infrastructure; 139
Murase Associates; 35
MVSA Architects; 123
MYS Architects; 155
National Hotels Co.; 154
NEK Beratende Ingenieure; 119
Net Group, The; 106
Nikken Sekkei, Ltd.; 92, 96, 109, 173
Nippon Building Fund, Inc.; 92
Nitsch Engineering; 39
Obayashi Corporation; 104, 105
Oculus Light Studio; 106
Odeh Engineers, Inc.; 39
Office of James Burnett, The; 35

Oger International; 139
OliverMcMillan; 53
OMA; 59, 94
omniCon Gesellschaft für innovatives Bauen mbH; 127
One Lux Studio, LLC; 139
Open Project; 134
Oppenheim Architecture + Design; 106
Orascom; 149
Orimoto Structural Engineers; 105
O'Sullivan Plumbing Inc.; 52
Parsons Brinkerhoff; 84
PBA; 106
PDW Architects; 104
Pei Cobb Freed & Partners; 100
Pelli Clarke Pelli Architects; 42, 104, 132
Pelton Marsh Kinsella; 149
Perkins + Will; 54
Persohn/Hahn Associates; 35
Philip Habib & Associates; 48
Philpotts Interiors; 53
Pickard Chilton; 35
PIESA; 53
Pitsou Kedem Architect; 145
Pivotal; 73
P Landscape Co., Ltd.; 107
Power; 44, 54
Priedemann Fassadenberatung GmbH; 127
Professional Services Industries; 35
Pronasa – Proterm; 54
Proycon; 55
Proyecta Ingenieros; 54
PT. Arkonin; 104
P & T Group; 107
PT. Karyadeka Graha Lestari; 104
PT. Kinematika; 104
PT. Malmass Mitra Teknik; 104
PT. Tata Mulia Nusantara; 104
Pulso Inmobiliario; 53
PVFC Land; 105
PWP Landscape Architecture; 100
Randal Brown and Assocaites Ltd.; 50
Renzo Piano Building Workshop; 113
Research Institute for Environmental Redevelopment; 105
Rider Levett Bucknall LLP; 69
RJA Group, Inc.; 102
RMJM; 135
Rolf Jensen & Associates, Inc.; 73
Rosenwasser/Grossman; 48
R.S. Cohen Safety Engineering, Ltd.; 154
Ruby+Associates; 44
Rush University Medical Center; 54
RWDI; 29, 44, 73, 102, 139, 159
RWG Associates; 159
Sahm-shin Engineers, Inc.; 88
Sako & Associates, Inc.; 44
Salomon Kamaji; 53
Salvador Aguilar; 46
Sansiri Public Company Limited; 107
Sauerbruch Hutton; 119
SEA Consult Engineering Co., Ltd.; 107
Sellar Property Group; 113
Serex International; 139
SGH; 39
Shanghai Construction Group; 73
Shanghai Institute of Architectural Design & Research; 65
Shanghai Liaoshen Curtain Wall Engineering Co., Ltd.; 102
Sharpes Redmore; 134
Shen Milsom Wilke, Inc.; 73, 84, 102
Shenzhen General Institute of Architectural Design and Research Co., Ltd.; 94, 108
Shenzhen King Façade; 82
Shenzhen Stock Exchange; 94
Shepherd Construction; 134
Siphya Construction Co., Ltd.; 107
Skanska AS; 159
Skidmore, Owings & Merrill LLP; 73, 102, 108, 181

SLCE Architects; 42
S. Netanel Engineers & Consultants, Ltd.; 145, 155
SNS Property Finance BV; 123
SOCSA; 52
Solomon Cordwell Buenz; 44
Southern Land Development Co., Ltd.; 100
Speirs and Major; 50
Steven Holl Architects; 77
Stiva; 46
Stonehenge Inter Co., Ltd.; 107
Student Castle; 134
Studio Leonardo Corbo; 132
Suffolk Construction; 39
SWA Group; 73, 102
Swiss Re; 159
Syntec Construction Public Co., Ltd.; 107
Taisei Corporation; 90, 109
Talon International Development, Inc.; 50
TCMC Architects & Engineers; 88
Teknik Yapi; 135
TEMA - Urban Landscape Design; 145
TEN Arquitectos; 48
TEP Consultants Pte Ltd.; 69
Terra Engineering; 54
Thomas Bell Wright International Consultants; 149
Thornton Tomasetti; 35, 54, 86, 139
TID Joint Stock Company; 105
Tierra Design Pte Ltd.; 69
Tiong Seng Contractors Pte Ltd.; 69
Tokyu Architects & Engineers, Inc.; 96, 105
Tokyu Construction; 96
Tongji Architectural Design (Group) Co., Ltd.; 82
Torres Paseo Colon S.A.; 55
Tosoni; 134
Transsolar; 29, 119
T.R.O.P., Ltd.; 107
Tuncay Akdag; 135
Turner & Townsend; 113
Two Trees Management; 48
U. Dori Group, Ltd.; 145, 155
Uni Engineering Co., Ltd.; 100
Unifimm; 134
Union Urban Redevelopment of Toranomon-Roppongi Area; 104
UOL Group Limited; 69
Urbanizadora Jardin S.A.; 54
V3 Companies of Illinois; 44
Varyap; 135
Verdaus; 152
Vertical Developments; 52
Vidal Arquitectos; 46, 52
Viridian Energy & Environmental, LLC; 48
Waxman Govrin Geva Engineering, Ltd.; 145, 155
Werner Sobek Group; 90, 119
William Vitacco Associates, Ltd.; 48
Windtech; 98
WOHA; 69
Wolff Landscape Architecture, Inc.; 52
Woowon M&E; 86
WSP Group; 39, 42, 113, 134, 139
Wuhan Lingyun; 82
Xiamen C&D Corporation Limited; 65
Xiamen C&D Real Estate Co., Ltd; 65
Yamashita Sekkei, Inc.; 104
Yixin Jinyuan Real Estate Co., Ltd.; 108
Yolles; 29, 50
Yungdo Engineers & Consultants; 88
ZdC; 46
Zeidler Partnership Architects; 29, 50
ZGF Architects LLP; 44
Zhejiang Construction Engineering Group Co., Ltd.; 82
Zhejiang Greatwall Construction Group Co., Ltd.; 82
Zhejiang Haitian Construction Group; 108
Zhejiang Zhong Tian Construction Group Co., Ltd.; 102
Zhongshan Shengxing; 82
Zur Wolf Landscape Architects; 155
Zvi Ronen Engineers, Ltd.; 155

Image Credits

Front Cover: (from left to right) The Bow, © Nigel Young/Foster + Partners; CCTV, © CCTV/Rem Koolhaas, former partner Ole Scheeren (until 2010), partner David Gianotten, photographed by Iwan Baan; The Shard, © Terri Meyer Boake; Sowwah Square, © Mubadala Real Estate & Infrastructure

- **Pg 8:** © Nigel Young/Foster + Partners
- **Pg 9:** © CCTV/Rem Koolhaas, former partner Ole Scheeren (until 2010), partner David Gianotten, photographed by Iwan Baan
- **Pg 10:** The Shard, © Tansri Muliani; Sowwah Square © Mubadala Real Estate & Infrastructure
- **Pg 11:** Tree House Residence Hall, © Chuck Choi; PARKROYAL on Pickering, © Patrick Bingham-Hall
- **Pg 12:** © Pearl River Tower Properties Co., Ltd.
- **Pg 13:** 30 St Mary Axe, © Steven Henry; BSB Construction Method, © BROAD Group
- **Pg 14:** © KONE
- **Pg 17–22:** All © CTBUH
- **Pg 28–33:** Images © Nigel Young/Foster + Partners; drawings © Foster + Partners
- **Pg 34–37:** Images © Alan Karchmer; drawings © Pickard Chilton
- **Pg 38:** © Peter Vanderwarker
- **Pg 39:** © Chuck Choi
- **Pg 40:** Image © Chuck Choi; drawing © ADD Inc
- **Pg 41:** All © Lucy Chen
- **Pg 42:** © Jeff Goldberg – Esto
- **Pg 43:** Top image © stefenturner.com; bottom image © Jeff Goldberg – Esto; drawings © Pelli Clarke Pelli Architects
- **Pg 44–45:** Images © Nick Merrick, Hedrich Blessing; drawing © ZGF Architects, LLP
- **Pg 46–47:** Images © Jorge Taboada; drawings © Vidal Arquitectos
- **Pg 48–49:** Images © Evan Joseph; drawing © Ten Arquitectos
- **Pg 50–51:** All © Zeidler Partnership Architects
- **Pg 52:** Coast at Lakeshore East, © bKL Architecture LLC / Darris Lee Harris Photography; Helicon, © Jorge Taboada
- **Pg 53:** Pacifica Honolulu, © Baldridge & Associates Structural Engineering, Inc.; Reforma 342 © Colonnier y Asociados
- **Pg 54:** Rush University Medical Center Hospital Tower, © Marshall Gerometta; Torre Betgonias, © Arquitectonica
- **Pg 55:** © Arquitectura y Diseño S.A.
- **Pg 58–59:** © OMA / Philippe Ruault
- **Pg 60:** Top image and drawing © OMA; bottom left image © OMA / Philippe Ruault
- **Pg 61:** © CCTV / Rem Koolhaas, former partner Ole Scheeren (until 2010), partner David Gianotten, photography by Iwan Baan
- **Pg 62:** © Arup
- **Pg 63:** Image © OMA / Jim Gourley; drawings © OMA
- **Pg 64–67:** All © Gravity Partnership, Ltd.
- **Pg 68–71:** Images © Patrick Bingham-Hall; drawing © WOHA
- **Pg 72:** © Pearl River Tower Properties Co., Ltd.
- **Pg 73:** © SOM
- **Pg 74:** © Pearl River Tower Properties Co., Ltd.
- **Pg 75:** All © SOM
- **Pg 76:** © Shu He
- **Pg 77:** © Steven Holl Architects
- **Pg 78:** Top image © Shu He; bottom image © Iwan Baan
- **Pg 79:** Image © Shu He; drawing © Steven Holl Architects
- **Pg 80–81:** Images © Peter Bennetts; drawing © HASSELL
- **Pg 82–83:** All © TJADRI
- **Pg 84–85:** Images © Tim Griffith & Grischa Ruschendorf; drawings © KPF
- **Pg 86:** © IFC Seoul
- **Pg 87:** Right image © Kitmin; left images © Namgoong, Sun; drawing © Arquitectonica
- **Pg 88–89:** All © Daewoo Engineering & Construction
- **Pg 90–91:** Images © Rainer Viertlboeck; drawing © JAHN
- **Pg 92–93:** All © Nikken Sekkei, Ltd.
- **Pg 94–95:** Images © OMA / Philippe Ruault; drawing © OMA
- **Pg 96–97:** All © Sstokyo
- **Pg 98:** © Tony Dowthwaite
- **Pg 99:** Images © Peter Sexty; drawing © DBI Design
- **Pg 100–101:** Images © Southern Land Development Co.; drawing © Pei Cobb Freed & Partners / J.M. Lin Architect
- **Pg 102–103:** Images © SOM / Si-ye Zhang; drawing © SOM
- **Pg 104:** Alamanda Office Tower, © PDW; ARK Hills Sengokuyama Mori Tower © Mori Building Co., Ltd.
- **Pg 105:** City Tower Kobe Sannomiya, © Tokyu Architects & Engineers, Inc.; Dolphin Plaza, © TID Joint Stock Company
- **Pg 106:** Huarun Tower, © Callison, LLC; Net Metropolis, © Adphoto
- **Pg 107:** All © Palmer & Turner (Thailand) Ltd.
- **Pg 108:** Shenzhen Kerry Plaza Phase II, © Kerry Development Co., Ltd.; Yixing Dongjiu, © ECADI
- **Pg 109:** © Nikken Sekkei, Ltd.
- **Pg 112:** © Terri Meyer Boake
- **Pg 113–114:** © Sellar Property Group 2013
- **Pg 115–116:** Top image © Terri Meyer Boake; bottom image © Steven Henry; drawings © RPBW
- **Pg 117:** © Sellar Property Group 2013
- **Pg 118:** © Jan Bitter
- **Pg 119:** © Stephan Liebl
- **Pg 120:** © Jan Bitter
- **Pg 121:** Image © Stephan Liebl; drawing © Sauerbruch Hutton
- **Pg 122:** © Jeroen Musch
- **Pg 123–124:** © Juriaan Brobbel
- **Pg 125:** Top left image © Mart Engelen; top right image © Rob Hoekstra; bottom left image © MVSA Architects; bottom right image © Jeroen Musch
- **Pg 126:** © Corinne Rose
- **Pg 127:** © Ina Reinecke / Barkow Leibinger
- **Pg 128:** © Barkow Leibinger
- **Pg 129:** All © Corinne Rose
- **Pg 130:** © Butyrskii Igor
- **Pg 131:** All © Liedel Investments
- **Pg 132:** © Pelli Clarke Pellie Architects
- **Pg 133:** Top left image © Dario Trabucco; bottom left image © Jessica Butler; right image (CC BY-SA 3.0) Walter J. Rotelmayer; drawing © Pelli Clarke Pelli Architects
- **Pg 134:** No. 1 Great Marlborough Street, © Daniel Hopkinson; Torre Unipol, © Open Project S.r.l.
- **Pg 135:** © Varyap
- **Pg 138:** © Lester Ali
- **Pg 139–140:** © Mubadala Real Estate & Infrastructure
- **Pg 141:** © Goettsch Partners
- **Pg 142:** Image © Lester Ali; drawing © Goettsch Partners
- **Pg 143:** All © Gerry O'Leary
- **Pg 144–147:** Images © Prof. Moshe Zur and Yael Engelhart; drawing © Moshe Tzur Architects and Town Planners, Ltd.
- **Pg 148–151:** All © Aldar Properties PJSC
- **Pg 152:** © JW Marriott Marquis
- **Pg 153:** Left images and drawing © Archgroup; right image © JW Marriott Marquis
- **Pg 154:** Diplomat Commercial Office Tower, © Chase Perdana Sdn Bhd.; Faire Tower, © Ronen Kook
- **Pg 155:** © Emily Paz-Tepper
- **Pg 158:** © Philip Oldfield
- **Pg 159:** © Nigel Young
- **Pg 160:** © James Newton
- **Pg 161:** All © Foster + Partners

Pg 162–163: Images © Nigel Young; drawings © Foster + Partners
Pg 164–167: All © BROAD Group
Pg 168–171: All © KONE
Pg 172–175: All © Nikken Sekkei, Ltd.
Pg 176: © Corinne Rose
Pg 177–179: All © Barkow Leibinger
180–183: All images © SOM / Tim Griffith; drawing © SOM
Pg 186: © Gorchev & Gorchev
Pg 187: © Emily Nemens, Center for Architecture
Pg 188: Top image © Associated Commercial Photographers; bottom image © Joseph Molitor, courtesy Avery Architectural and Fine Arts Library, Columbia University
Pg 189: Top image © Norman McGrath; bottom image © Wes Thompson
Pg 190: Left top and bottom images © Gérald Morand; right top and bottom images © Fernando Guerra I FG + SG
Pg 191: All © Fernando Guerra I FG + SG
Pg 192: © Marshall Gerometta
Pg 194–195: All © Marshall Gerometta
Pg 196: © Taipei Financial
Pg 197: © Tansri Muliani
Pg 200–201: All © CTBUH
Pg 202: Top image © Tom Arban; bottom image © CTBUH
Pg 203: Left image © CTBUH; right image © H.G. Esch, Hennef
Pg 204: Top image © Fernando Guerra; bottom image © CTBUH
Pg 205: Top left image © Aedas; bottom left and top right images © CTBUH; bottom right image © Ateliers Jean Nouvel
Pg 207: Beetham Hilton Tower, © Ian Simpson Architects; Hearst Tower, © Antony Wood/CTBUH; Shanghai World Financial Center, © Mori Building Co., Ltd.; New York Times Building, © David Sundberg/Esto; 51 Lime Street, © Nigel Young/ Foster + Partners; Bahrain World Trade Center, © Atkins; Linked Hybrid, © Shu He; Manitoba Hydro Place, © Smith Carter; The Broadgate Tower, SOM | Richard Leeney © British Land; Tornado Tower, © Grey Doha; Burj Khalifa, SOM | Nick Merrick © Hedrich Blessing 2010; Broadcasting Place, © Sapa Architectural Services; Bank of America Tower, © David Sundberg/Esto; Pinnacle @ Duxton, © ARC Studio Architecture + Urbanism and RSP Architects Planners & Engineers (Pte) Ltd.; KfW Westarkade, © Jan Bitter; Eight Spruce Street, © Marshall Gerometta; Guangzhou International Finance Centre, © Christian Richters; The Index, © Nigel Young/Foster + Partners; Doha Tower, © Ateliers Jean Nouvel; Absolute Towers, © Tom Arban; 1 Bligh Street, © H.G. Esch, Hennef; Palazzo Lombardia, © Fernando Guerra; Al Bahar Towers, © Aedas
Pg 208–209: © CTBUH

CTBUH Organizational Structure & Members

Board of Trustees

Chairman: Timothy Johnson, NBBJ, USA
Executive Director: Antony Wood, CTBUH & Illinois Institute of Technology, USA
Secretary: William Maibusch, Turner Construction International LLC, Qatar
Treasurer: Steve Watts, AECOM/Davis Langdon LLP, UK
Trustee: William Baker, Skidmore, Owings & Merrill LLP, USA
Trustee: Craig Gibbons, Arup, Australia
Trustee: David Malott, Kohn Pedersen Fox, USA
Trustee: Dennis Poon, Thornton Tomasetti, USA
Trustee: Cathy Yang, Taipei Financial Center Corporation, Taiwan, China

Staff / Contributors

Executive Director: Antony Wood
Operations: Patti Thurmond
Office Coordinator / Bookkeeper: Christy Zvonar
Membership: Carissa Devereux
Editor: Daniel Safarik
Design & Production: Steven Henry
Research Associate: Payam Bahrami
Research Associate: Dario Trabucco
Graphics Production Associate: Marty Carver
Database Editor: Marshall Gerometta
Publications Associate: Tansri Muliani
Web Developer: Son Dang
Website Editor: Aric Austermann
General Counsel: Joseph Dennis
CTBUH Journal Associate Editor: Robert Lau
Special Media Correspondent: Jeff Herzer

Advisory Group

Ahmad K. Abdelrazaq, Samsung Corporation, Korea
Dimitrios Antzoulis, Turner International LLC, USA
Carl Baldassarra, Rolf Jensen Associates, USA
Joseph G. Burns, Thornton Tomasetti, USA
Johannes de Jong, KONE International, Finland
Mahjoub Elnimeiri, Illinois Institute of Technology, USA
Thomas K. Fridstein, AECOM Enterprises, USA
Mark J. Frisch, Solomon Cordwell Buenz, USA
Mayank Gandhi, DK Infrastructure, India
Paul James, Lend Lease, USA
Charles Killebrew, John Portman & Associates, USA
Simon Lay, AECOM, UK
Moira M. Moser, M. Moser Associates, Hong Kong
John Nipaver, John Portman & Associates, USA
Jerry R. Reich, Chicago Committee on High-rise Buildings, Inc., USA
Mark P. Sarkisian, Skidmore, Owings & Merrill LLP, USA
David Scott, Laing O'Rourke, UK
Brett Taylor, Bornhorst + Ward Consulting Engineers, Australia

Working Group Co-Chairs

Façade Access: Lance McMasters, Kevin Thompson, Lee Herzog & Peter Weismantle
Finance & Economics: Steve Watts
Fire & Safety: Jose L. Torero & Daniel O' Connor
Foundations: Frances Badelow, Tony Kiefer, Sungho Kim, James Sze, George Leventis & Rudy Frizzi
Legal Aspects of Tall Buildings: Cecily Davis
Outrigger Systems: Hi Sun Choi, Leonard Joseph, Neville Mathias & Goman Ho
Research, Academic & Postgraduate: Philip Oldfield & Dario Trabucco
Seismic Design: Ron Klemencic, Andrew Whittaker & Michael Willford
Sustainable Design: Antony Wood
Wind Engineering: Peter Irwin, Roy Denoon, David Scott

Regional Representatives

Australia: Bruce Wolfe, Conrad Gargett Architecture
Austria: Matthäus Groh, KS Ingenieure ZT GmbH
Belarus: George Shpuntov, JV "Aexandrov-Passage" LLC
Belgium: Georges Binder, Buildings & Data S.A.
Brazil: Antonio Macêdo Filho, EcoBuilding Consulting
Canada: Barry Charnish, Entuitive
China: Guo-Qiang Li, Tongji Univesity
Costa Rica: Ronald Steinvorth, IECA International
Finland: Mikko Korte, KONE Corporation
Germany: Werner Sobek, Werner Sobek Stuttgart GmbH & Co.
Greece: Alexios Vandoros, Vandoros & Partners
Hong Kong: Stefan Krummeck, TFP Farrells
India: Mayank Gandhi, DK Infrastructure
Indonesia: Tiyok Prasetyoadi, PDW Architects
Iran: Peyman Askarinejad, TJEG International
Israel: Israel David, David Engineers
Italy: Dario Trabucco, IUAV University of Venice
Japan: Masayoshi Nakai, Takenaka Corporation
Lebanon: Ramy El-Khoury, Rafik El-Khoury & Partners
Mexico: Ricardo Nankin, Grupo Elipse
New Zealand: Simon Longuet-Higgins, Beca Group
Philippines: Felino A. Palafox, Palafox Associates
Poland: Ryszard M. Kowalczyk, University of Beira Interior
Qatar: William Maibusch, Turner Construction International
Romania: Mihail Iancovici, Technical University of Civil Engineering of Bucharest (UTCB)
Russia: Elena Shuvalova, Lobby Agency
Saudi Arabia: Bassam Al-Bassam, Rayadah Investment Company, KSA
Singapore: Juneid Qureshi, Meinhardt (S) Pte Ltd.
South Africa: Kevan Moses, Stauch Vorster Architects
South Korea: Dr. Kwang Ryang Chung, Dongyang Structural Engineers Co., Ltd
Spain: Javier Quintana De Uña, Taller Básico de Arquitectura SLP
Taiwan: Cathy Yang, Taipei Financial Center Corp.
Thailand: Pennung Warnitchai, Asian Institute of Technology
Turkey: Can Karayel, Langan International
United Kingdom: Steve Watts, AECOM/Davis Langdon LLP
Vietnam: Phan Quang Minh, National University of Civil Engineering

Committee Chairs

Awards: Jeanne Gang, Studio Gang, USA
Height: Peter Weismantle, Adrian Smith + Gordon Gill Architecture, USA
Young Professionals: Sergio Valentini, JAHN, USA

CTBUH Organizational Members

(as of July 2013) http://membership.ctbuh.org

Supporting Contributors

AECOM
Al Hamra Real Estate Company
Broad Sustainable Building Co., Ltd.
BT – Applied Technology
Buro Happold, Ltd.
China State Construction Engineering Corporation (CSCEC)
CITIC HEYE Investment (Beijing) Co., Ltd.
Dow Corning Corporation
Emaar Properties, PJSC
Eton Properties (Dalian) Co., Ltd.
HOK, Inc.
Illinois Institute of Technology
Kingdom Real Estate Development Co.
Kohn Pedersen Fox Associates, PC
KONE Industrial, Ltd.
Lotte Engineering & Construction Co.
Morin Khuur Tower LLC
NBBJ
Permasteelisa Group
Samsung C&T Corporation
Shanghai Tower Construction & Development Co., Ltd.
Skidmore, Owings & Merrill LLP
Taipei Financial Center Corp. (TAIPEI 101)
Turner Construction Company
Underwriters Laboratories (UL) LLC
Woods Bagot
WSP Group

Patrons

Akzo Nobel
Al Ghurair Construction – Aluminum LLC
Arabtec Construction LLC
Blume Foundation
BMT Fluid Mechanics, Ltd.
Durst Organization
East China Architectural Design & Research Institute Co., Ltd.
Gensler
Hongkong Land, Ltd.
KLCC Property Holdings Berhad
Langan Engineering & Environmental Services, Inc.
Meinhardt Group International
Schindler Elevator Corporation
Shanghai Institute of Architectural Design & Research Co., Ltd.
Studio Daniel Libeskind
Thornton Tomasetti, Inc.
Tishman Speyer Properties
Weidlinger Associates, Inc.
Zuhair Fayez Partnership

Donors

Adrian Smith + Gordon Gill Architecture, LLP
American Institute of Steel Construction
Aon Fire Protection Engineering Corp.
ARCADIS/The Rise Group LLC
Arup
Aurecon
Besix SA
Bollinger + Grohmann Ingenieure
Brookfield Multiplex Construction Europe, Ltd.
C.Y. Lee & Partners Architects/Planners
CCDI Group
Enclos Corp.
Fender Katsalidis
Halfen USA
Hyundai Steel Company
Jacobs
Laing O'Rourke
Larsen & Toubro, Ltd.
Leslie E. Robertson Associates, RLLP
Magnusson Klemencic Associates, Inc.
Maire Tecnimont Group
MAKE
Mooyoung Architects & Engineers
MulvannyG2 Architecture
Nishkian Menninger Consulting and Structural Engineers
PDW Architects
Pei Cobb Freed & Partners
Pick ard Chilton Architects, Inc.
PT Gistama Intisemesta
Quadrangle Architects Ltd.
Rafik El-Khoury & Partners
Rolf Jensen & Associates, Inc.
Rowan Williams Davies & Irwin, Inc.
RTKL Associates Inc.
Saudi Binladin Group / ABC Division
Severud Associates Consulting Engineers, PC

Shanghai Construction (Group) General Co., Ltd.
Shanghai Jingang North Bund Realty Co., Ltd (Subsidiary of Sinar Mas Group – APP China)
Shree Ram Urban Infrastructure, Ltd.
SIAPLAN Architects and Planners
Skanska
Solomon Cordwell Buenz
Studio Gang Architects
SWA Group
Syska Hennessy Group, Inc.
T.Y. Lin International Pte. Ltd.
Tongji Architectural Design (Group) Co., Ltd.
Walter P. Moore and Associates, Inc.
Werner Voss + Partner
Yolles

Contributors

Aedas
Allford Hall Monaghan Morris Ltd.
Alvine Engineering
Barker Mohandas, LLC
Bates Smart
Benoy Limited
Bonacci Group
Boundary Layer Wind Tunnel Laboratory
Bouygues Construction
British Land Company PLC
C S Structural Engineering, Inc.
Canary Wharf Group, PLC
Canderel Management, Inc.
CBRE Group, Inc.
CCL Qatar w.l.l.
Continental Automated Buildings Association
DBI Design Pty Ltd
DCA Architects Pte Ltd
Deerns Consulting Engineers
DK Infrastructure Pvt. Ltd.
DongYang Structural Engineers Co., Ltd.
Ellumus LLC
Far East Aluminium Works Co., Ltd.
GGLO, LLC
Goettsch Partners
Gradient Microclimate Engineering Inc. (GmE)
Graziani + Corazza Architects Inc.
HAEAHN Architecture, Inc.
Hariri Pontarini Architects
Hiranandani Group
Hyder Consulting (Shanghai)
Israeli Association of Construction and Infrastructure Engineers
J. J. Pan and Partners, Architects and Planners
Jiang Architects & Engineers
KHP Konig und Heunisch Planungsgesellschaft
Langdon & Seah Singapore
Lend Lease
Liberty Group Properties
M Moser Associates Ltd.
Mori Building Co., Ltd.
Nabih Youssef & Associates
National Fire Protection Association
National Institute of Standards and Technology
Norman Disney & Young
Ornamental Metal Institute of New York
Otis Elevator Company
Parsons Brinckerhoff
Perkins + Will
Pomeroy Studio Pte Ltd
PositivEnergy Practice, LLC
RAW Design Inc.
Ronald Lu & Partners
Rosenwasser/Grossman Consulting Engineers, PC
Royal HaskoningDHV
Samcon Gestion Inc.
SAMOO Architects & Engineers
Sanni, Ojo & Partners
Silvercup Studios
SilverEdge Systems Software, Inc.
SIP Project Managers Pty Ltd
Steel Institute of New York
Tekla Corporation
Terrell Group
ThyssenKrupp Elevator
TSNIIEP for Residential and Public Buildings
University of Illinois at Urbana-Champaign
Vetrocare SRL
Wilkinson Eyre Architects

Participants

ACSI (Ayling Consulting Services Inc)
Adamson Associates Architects
Aidea Philippines, Inc.
AKF Group, LLC
Al Jazera Consultants
ALT Cladding, Inc.
ARC Studio Architecture + Urbanism
ArcelorMittal
Architects 61 Pte., Ltd.
Architectural Design & Research Institute of Tsinghua University
Architectural Institute of Korea
Architectus
Arquitectonica International Corp.
ASL Sencorp
Atkins
Azrieli Group Ltd.
Bakkala Consulting Engineers Limited
BAUM Architects
BDSP Partnership
Beca Group
Benchmark
BG&E Pty., Ltd.
BIAD (Beijing Institute of Architectural Design)
Bigen Africa Services (Pty) Ltd.
Billings Design Associates, Ltd.
bKL Architecture LLC
BluEnt
Boston Properties, Inc.
Broadway Malyan
Callison, LLC
Camara Consultores Arquitectura e Ingenieria
Capital Group
Case Foundation Co.
CB Engineers
CCHRB (Chicago Committee on High-Rise Buildings)
CDC Curtain Wall Design & Consulting, Inc.
Central Scientific and Research Institute of Engineering Structures "SRC Construction"
Cermak Peterka Petersen, Inc. (CPP Inc.)
China Academy of Building Research
China Institute of Building Standard Design & Research (CIBSDR)
Chinachem Group
City Developments Limited
Code Consultants, Inc.
Concrete Reinforcing Steel Institute (CRSI)
COOKFOX Architects
Cosentini Associates
COWI A/S
Cox Architecture Pty. Ltd.
CS Associates, Inc.
CTL Group
Cundall
Dar Al-Handasah (Shair & Partners)
Degracuwe Consulting
Delft University of Technology
Dennis Lau & Ng Chun Man Architects & Engineers (HK), Ltd.
dhk Architects Pty., Ltd.
Diar Consult
DSP Design Associates Pvt., Ltd.
Dunbar & Boardman
ECSD S.r.l.
Edgett Williams Consulting Group, Inc.
Edmonds International USA
Eight Partnership Ltd.
Electra Construction LTD
ENAR, Envolventes Arquitectonicas
Ennead Architects LLP
Environmental Systems Design, Inc.
Epstein
Façade India Testing Inc.
Feilden Clegg Bradley Studios LLP
Fortune Shepler Consulting
FXFOWLE Architects, LLP
Gale International / New Songdo International City Development, LLC
GCAQ Ingenieros Civiles S.A.C.
GEO Global Engineering Consultants
Gilsanz Murray Steficek
Glass Wall Systems (India) Pvt. Ltd
Gold Coast City Council
Gorprojekt (Urban Planning Institute of Residential and Public Buildings)
Grace Construction Products
Greyling Insurance Brokerage
Grimshaw Architects
Grupo Inmobiliario del Parque
Guangzhou Scientific Computing Consultants Co., Ltd.
GVK Elevator Consulting Services, Inc.
Halvorson and Partners
Haynes-Whaley Associates, Inc.
Heller Manus Architects
Henning Larsen Architects
Hilson Moran Partnership, Ltd.
Hong Kong Housing Authority
Hong Kong Polytechnic University
Housing and Development Board
IECA Internacional S.A.
ingenhoven architects
Institute BelNIIS, RUE
INTEMAC, SA
Irwinconsult Pty., Ltd.
Iv-Consult b.v.
Jahn, LLC
Jaros Baum & Bolles
Jaspers-Eyers Architects
JBA Consulting Engineers, Inc.
JCE Structural Engineering Group, Inc.
JMB Realty Corporation
John Portman & Associates, Inc.
Johnson Pilton Walker Pty. Ltd.
JV "Alexandrov-Passage" LLC
Kalpataru Limited
KEO International Consultants
Kinetica
King Saud University College of Architecture & Planning
KPFF Consulting Engineers
KPMB Architects
LCL Builds Corporation
Leigh & Orange, Ltd.
Lerch Bates, Inc.
Lerch Bates, Ltd. Europe
Living Architecture Inc.
Lobby Agency
Louie International Structural Engineers
Mace Limited
MADY
Magellan Development Group, LLC
Manitoba Hydro
Margolin Bros. Engineering & Consulting, Ltd.
McHugh Construction Co.
McNamara / Salvia, Inc.
Meinhardt (Thailand) Ltd.
Michael Blades & Associates
MKPL Architects Pte Ltd
Moshe Tzur Architects Town Planners Ltd.
New World Development Company Limited
Nikken Sekkei, Ltd.
Novawest LLC
NPO SODIS
O'Connor Sutton Cronin
Option One International, WLL
P&T Group
Palafox Associates
Paragon International Insurance Brokers Ltd.
Pelli Clarke Pelli Architects
PLP Architecture
PPG Industries, Inc.
Profica Project Management
Project and Design Research Institute "Novosibirsky Promstroyproject"
Rafael Viñoly Architects, PC
Read Jones Christoffersen Ltd.
Rene Lagos Engineers
Riggio / Boron, Ltd.
Roosevelt University – Marshall Bennett Institute of Real Estate
Sauerbruch Hutton Verwaltungsges mBH
schlaich bergermann und partner
Schock USA Inc.
Sematic SPA
Shimizu Corporation
SKS Associates
SMDP, LLC
SmithGroup
Southern Land Development Co., Ltd.
St. Francis Square Development Corp.
Stanley D. Lindsey & Associates, Ltd.
Stauch Vorster Architects
Stephan Reinke Architects, Ltd.
Studio Altieri S.p.A.
Sufrin Group
Taisei Corporation
Takenaka Corporation
Tameer Holding Investment LLC
Tandem Architects (2001) Ltd.
Taylor Thomson Whitting Pty., Ltd.
TFP Farrells, Ltd.
Thermafiber, Inc.
Transsolar
Trump Organization
Tsao & McKown Architects, P.C.
Tyréns
University of Maryland – Architecture Library
University of Nottingham
UralNIIProject RAACS
Vanguard Realty Pvt., Ltd.
VDA (Van Deusen & Associates)
Vipac Engineers & Scientists, Ltd.
VOA Associates, Inc.
Walsh Construction Company
Weiss Architects, LLC
Werner Sobek Stuttgart GmbH & Co., KG
wh-p GmbH Beratende Ingenieure
Windtech Consultants Pty., Ltd.
WOHA Architects Pte., Ltd.
Wong & Ouyang (HK), Ltd.
Wordsearch
World Academy of Science for Complex Safety
WTM Engineers International GmbH
WZMH Architects
Y. A. Yashar Architects
Ziegler Cooper Architects

Supporting Contributors are those who contribute $10,000;
Patrons: $6,000; Donors: $3,000; Contributors: $1,500; Participants: $750